Kernel Methods and Hybrid Evolutionary Algorithms in Energy Forecasting

Kernel Methods and Hybrid Evolutionary Algorithms in Energy Forecasting

Special Issue Editor

Wei-Chiang Hong

MDPI • Basel • Beijing • Wuhan • Barcelona • Belgrade

MDPI

Special Issue Editor
Wei-Chiang Hong
School of Computer Science and Technology,
Jiangsu Normal University
China

Editorial Office
MDPI
St. Alban-Anlage 66
Basel, Switzerland

This is a reprint of articles from the Special Issue published online in the open access journal *Energies* (ISSN 1996-1073) in 2016 (available at: https://www.mdpi.com/journal/energies/special_issues/algorithms-energy-forecasting)

For citation purposes, cite each article independently as indicated on the article page online and as indicated below:

LastName, A.A.; LastName, B.B.; LastName, C.C. Article Title. *Journal Name* **Year**, *Article Number*, Page Range.

ISBN 978-3-03897-292-1 (Pbk)
ISBN 978-3-03897-293-8 (PDF)

Contents

About the Special Issue Editor

Wei-Chiang Hong, Jiangsu Distinguished Professor, School of Computer Science and Technology, Jiangsu Normal University, China. His research interests mainly include computational intelligence (neural networks, evolutionary computation) and application of forecasting technology (ARIMA, support vector regression, and chaos theory). In May 2012, his paper was evaluated as a "Top Cited Article 2007–2011" by *Applied Mathematical Modelling* (Elsevier). In August 2014, he was elected to be awarded the "Outstanding Professor Award" by the Far Eastern Y. Z. Hsu Science and Technology Memorial Foundation (Taiwan). In November 2014, he was elected to be awarded the "Taiwan Inaugural Scopus Young Researcher Award—Computer Science", by Elsevier, in the Presidents' Forum of Southeast and South Asia and Taiwan Universities. He was awarded as one of the "Top 10 Best Reviewers" of Applied Energy in 2014, and as one of Applied Energy's "Best Reviewers" in 2016.

Preface to "Kernel Methods and Hybrid Evolutionary Algorithms in Energy Forecasting"

The development of kernel methods and hybrid evolutionary algorithms (HEAs) to support experts in energy forecasting is of great importance to improving the accuracy of the actions derived from an energy decision maker, and it is crucial that they are theoretically sound. In addition, more accurate or more precise energy demand forecasts are required when decisions are made in a competitive environment. Therefore, this is of special relevance in the Big Data era. These forecasts are usually based on a complex function combination. These models have resulted in over-reliance on the use of informal judgment and higher expense if lacking the ability to catch the data patterns. The novel applications of kernel methods and hybrid evolutionary algorithms can provide more satisfactory parameters in forecasting models.

This book contains articles from the Special Issue titled "Kernel Methods and Hybrid Evolutionary Algorithms in Energy Forecasting", which aimed to attract researchers with an interest in the research areas described above. As Fan et al. [1] indicate, the research direction of energy forecasting in recent years has concentrated on proposing hybrid or combined models: (1) hybridizing or combining these artificial intelligence models with each other; (2) hybridizing or combining with traditional statistical tools; and (3) hybridizing or combining with those superior evolutionary algorithms. Therefore, this Special Issue sought contributions towards the development of HEAs with kernel methods or with other novel methods (e.g., chaotic mapping mechanism, fuzzy theory, and quantum computing mechanism), which, with superior capabilities over the traditional optimization approaches, aim to overcome some embedded drawbacks and then apply these new HEAs to be hybridized with original forecasting models to significantly improve forecasting accuracy.

The 10 articles collected in this compendium all display a broad range of cutting-edge topics in the kernel methods and hybrid evolutionary algorithms. The preface author believes that these hybrid approaches will play an important role in energy forecasting accuracy improvements. It is known that the evolutionary algorithms have their theoretical drawbacks, such as a lack of knowledge, memory, or storage functions; they are time consuming in training; and become trapped in local optima. Therefore, the goal of hybridizing optimization methods to adjust their internal parameters (e.g., mutation rate, crossover rate, annealing temperature, etc.) is to overcome these shortcomings. Firstly, for example, in genetic algorithms (GAs), new individuals are generated by the following operators: selection, crossover, and mutation. For all types of objective functions, the generation begins with a binary coding for the parameter set. Based on this special binary coding process, GAs are able to solve some specified problems which are not easily solved by traditional algorithms. GAs can empirically provide a few best-fitted offspring from the whole population, but after several generations, due to low population diversity, it might lead to a premature convergence. Due to the easy implementation process and a special mechanism to escape from local optima, chaos and chaos-based searching algorithms have received intense attention. Applications of chaotic mapping mechanisms to carefully expand variable searching space (i.e., allow variables to travel ergodically over the search space) are increasingly popularly employed in evolutionary computation fields.

Secondly, several disadvantages embedded in these evolutionary algorithms need to be improved in order to achieve a more satisfactory performance. For example, based on the operation procedure of simulated annealing algorithm (SA), subtle and skillful adjustment in the annealing schedule is required, such as the size of the temperature steps during annealing. Particularly, the temperature of each state is discrete and unchangeable, which does not meet the requirement of continuous decrease in temperature in actual physical annealing processes.

In addition, SA easily accepts deteriorated solutions with high temperature, and it is difficult to escape from local minimum traps at low temperature. Cloud theory is considered to overcome these drawbacks, as demonstrated in Geng et al. [2]. Cloud theory is a model of the uncertainty transformation between quantitative representation and qualitative concept using language value. Based on the SA operation procedure, subtle and skillful adjustment in the annealing schedule is required (e.g., the size of the temperature steps during annealing, the temperature range, the number of re-starts and re-direction of the search). The annealing process is like a fuzzy system in which the molecules move from large-scale to small-scale randomly as the temperature decreases. In addition, due to its Monte Carlo scheme and lack of knowledge memory functions, its time-consuming nature is another problem. Geng et al. [2] tried to employ a chaotic simulated annealing (CSA) algorithm to overcome these shortcomings. In this, the transiently chaotic dynamics are temporarily generated for foraging and self-organizing. They are then gradually vanished with autonomous decrease of the temperature, and are accompanied by successive bifurcations and converged to a stable equilibrium. Therefore, CSA significantly improves the randomization of the Monte Carlo scheme, and controlled the convergent process by bifurcation structures instead of stochastic "thermal" fluctuations, eventually performing efficient searching including a global optimum state. However, as mentioned above, the temperature of each state is discrete and unchangeable, which does not meet the requirement of continuous decrease in temperature in actual physical annealing processes. Even if some temperature annealing functions are exponential in general, the temperature gradually falls with a fixed value in every annealing step and the changing process of temperature between two neighbor steps is not continuous. This phenomenon also appears when other types of temperature update functions are implemented (e.g., arithmetical, geometrical, or logarithmic). In cloud theory, by introducing the Y condition normal cloud generator to the temperature generation process, it can randomly generate a group of new values that distribute around the given value like a "cloud". The fixed temperature point of each step becomes a changeable temperature zone in which the temperature of each state generation in every annealing step is chosen randomly, the course of temperature change in the whole annealing process is nearly continuous, and fits the physical annealing process better. Therefore, based on chaotic sequence and cloud theory, the chaotic cloud simulated annealing algorithm (CCSA) is employed to replace the stochastic "thermal" fluctuations control from traditional SA to enhance the continuous physical temperature annealing process from CSA. Cloud theory can realize the transformation between a qualitative concept in words and its numerical representation. It can be employed to avoid the problems mentioned above.

Thirdly, the concepts of combined or hybrid models also deserve consideration. Note that the term "hybrid" means that some process of the former model is integrated into the process of the later one. For example, hybrid A and B implies some processes of A are controlled by A, and some are controlled by B. On the other hand, for the so-called combined models, the output of the former model becomes the input of the latter one. Therefore, the classification results from combined models will be superior to a single model. Combined models are employed to further capture more data pattern information from the analyzed data series. For example, inspired by the concept of recurrent neural networks (RNNs) where every unit is considered as an output of the network and the provision of adjusted information as input in a training process, the recurrent learning mechanism framework is also combined into the original analyzed model. For a feed-forward neural network, links can be established within layers of a neural network. These types of networks are called recurrent neural networks. RNNs are extensively applied in time series forecasting. Jordan [3] proposes a recurrent neural network model for controlling robots. Elman [4] develops a recurrent neural network model to solve linguistics problems. Williams and Zipser [5] present a recurrent network model to solve nonlinear adaptive filtering and pattern recognition problems. These three models mentioned all consist of a multilayer perceptron (MLP) with a hidden layer. Jordan's networks have a feedback loop from the output layer with past values to an additional input, namely a "context layer". Then, output values from the context layer are fed back into the hidden layer. Elman's networks have a feedback loop from the hidden layer to the context layer. In the networks of Williams and Zipser, nodes in the hidden layer are fully connected to each other. Both Jordan's and Elman's networks include an additional information source from the output layer or the hidden layer. Hence, these models use mainly past information to capture detailed information. The networks of Williams and Zipser take much more information from the hidden layer and feed it back into themselves. Therefore, the networks of Williams and Zipser are sensitive when models are implemented. On the other hand, for another combined model, some data series sometimes reveal a seasonal tendency due to cyclic economic activities or seasonal nature hour to hour, day to day, week to week, month to month, and season to season, such as an hourly peak in a working day, a weekly peak in a business week, and a monthly peak in a demand-planned year. In order to excellently deal with cyclic/seasonal trend data series, some useful trial (e.g., seasonal mechanism) is also received some intentions. The preface author proposed a seasonal mechanism [2,6,7] with two steps for convenience in implementation: the first step is calculating the seasonal index (SI) for each cyclic point in a cycle length peak period; the second step is computing the forecasting value by multiplying the seasonal index (SI).

This discussion of the work by the author of this preface highlights work in an emerging area of kernel methods and hybrid evolutionary algorithms that has come to the forefront over the past decade. The articles in this collection span many cutting-edge areas that are truly interdisciplinary in nature.

Wei-Chiang Hong
Guest Editor

Reference

1. Fan, G.F.; Peng, L.L.; Hong, W.C. Short term load forecasting based on phase space reconstruction algorithm and bi-square kernel regression model. *Appl. Energy* **2018**, *224*, 13–33.

2. Geng, J.; Huang, M.L.; Li, M.W.; Hong, W.C. Hybridization of seasonal chaotic cloud simulated annealing algorithm in a SVR-based load forecasting model. *Neurocomputing* **2015**, *151*, 1362–1373.

3. Jordan, M.I. Attractor dynamics and parallelism in a connectionist sequential machine. In Proceedings of the 8th Annual Conference of the Cognitive Science Society, Englewood Cliffs, NJ, USA, 1987, pp. 531–546.

4. Elman, J.L. Finding structure in time. *Cogn. Sci.* **1990**, *14*, 179–211.

5. Williams, R.; Zipser, D. A learning algorithm for continually running fully recurrent neural networks. *Neural Comput.* **1989**, *1*, 270–280.

6. Hong, W.C.; Dong, Y.; Zhang, W.Y.; Chen, L.Y.; Panigrahi, B.K. Cyclic electric load forecasting by seasonal SVR with chaotic genetic algorithm. *Int. J. Electr. Power Energy Syst.* **2013**, *44*, 604–614.

7. Ju, F.Y.; Hong, W.C. Application of seasonal SVR with chaotic gravitational search algorithm in electricity forecasting. *Appl. Math. Modelling* **2013**, *37*, 9643–9651.

energies

MDPI

Article

Hybridizing DEMD and Quantum PSO with SVR in Electric Load Forecasting

Li-Ling Peng [1], Guo-Feng Fan [1], Min-Liang Huang [2] and Wei-Chiang Hong [3,4,*]

[1]   College of Mathematics & Information Science, Ping Ding Shan University, Pingdingshan 467000, China; plling1054@163.com (L.-L.P.); guofengtongzhi@163.com (G.-F.F.)
[2]   Department of Industrial Management, Oriental Institute of Technology, 58 Sec. 2, Sichuan Rd., Panchiao, New Taipei 220, Taiwan; minglianghuang2016@gmail.com
[3]   School of Economics & Management, Nanjing Tech University, Nanjing 211800, China
[4]   Department of Information Management, Oriental Institute of Technology, 58 Sec. 2, Sichuan Rd., Panchiao, New Taipei 220, Taiwan
*   Correspondence: fi013@mail.oit.edu.tw, Tel.: +886-2-7738-0145 (ext. 5316); Fax: +886-277-386-310

Academic Editor: Sukanta Basu
Received: 5 February 2016; Accepted: 16 March 2016; Published: 19 March 2016

Abstract: Electric load forecasting is an important issue for a power utility, associated with the management of daily operations such as energy transfer scheduling, unit commitment, and load dispatch. Inspired by strong non-linear learning capability of support vector regression (SVR), this paper presents an SVR model hybridized with the differential empirical mode decomposition (DEMD) method and quantum particle swarm optimization algorithm (QPSO) for electric load forecasting. The DEMD method is employed to decompose the electric load to several detail parts associated with high frequencies (intrinsic mode function—IMF) and an approximate part associated with low frequencies. Hybridized with quantum theory to enhance particle searching performance, the so-called QPSO is used to optimize the parameters of SVR. The electric load data of the New South Wales (Sydney, Australia) market and the New York Independent System Operator (NYISO, New York, USA) are used for comparing the forecasting performances of different forecasting models. The results illustrate the validity of the idea that the proposed model can simultaneously provide forecasting with good accuracy and interpretability.

Keywords: electric load forecasting; support vector regression; quantum theory; particle swarm optimization; differential empirical mode decomposition; auto regression

1. Introduction

Electric energy can not be reserved, thus, electric load forecasting plays a vital role in the daily operational management of a power utility, such as energy transfer scheduling, unit commitment, load dispatch, and so on. With the emergence of load management strategies, it is highly desirable to develop accurate, fast, simple, robust and interpretable load forecasting models for these electric utilities to achieve the purposes of higher reliability and better management [1].

In the past decades, researchers have proposed lots of methodologies to improve the load forecasting accuracy level. For example, Bianco *et al.* [2] proposed linear regression models for electricity consumption forecasting; Zhou *et al.* [3] applied a Grey prediction model for energy consumption; Afshar and Bigdeli [4] presented an improved singular spectral analysis method to predict short-term load in the Iranian power market; and Kumar and Jain [5] compared the forecasting performances among three Grey theory-based time series models to explore the consumption situation of conventional energy in India. Bianco *et al.* [6] indicate that their load model could be successfully used as an input of broader models than those of their previous paper [2]. References [7–10] proposed several useful artificial neural networks models to conduct short-term load forecasting. The authors of [11–14] proposed hybrid models with evolutionary algorithms that demonstrated improved energy forecasting performances. These methods can achieve significant improvements in terms of forecasting accuracy, but without reasonable interpretability, particularly for ANN models. Artificial neural networks (ANNs), with mature nonlinear mapping capabilities and data processing characteristics, have achieved widely successful applications in load forecasting. Recently, expert systems with fuzzy rule-based linguistic means provided good interpretability while dealing with system modeling [15]. Various approaches and models have been proposed in the last decades in many area such as climate factors (temperature and humidity), social activities (human social activities), seasonal factors (seasonal climate change and load growth), and so on. However, these models have strong dependency on an expert and lack expected forecasting accuracy. Therefore, combination models which are based on these popular methods and other techniques can satisfy the two desired requests: high accuracy level and interpretability.

With superiority in handling high dimension nonlinear data, support vector regression (SVR) has been successfully used to solve forecasting problems in many fields, such as financial time series (stocks index and exchange rate) forecasting, tourist arrival forecasting, atmospheric science forecasting, and so on [16–18]. However, SVR methods have a significant disadvantage, in that while its three parameters are determined simultaneously during the nonlinear optimization process, the solution is easily trapped into a local optimum. In addition, it also lacks a statistically significant level of robustness. These two shortcomings are the focused topics in the SVR research field [19]. On the other hand, the empirical mode decomposition (EMD) with auto regression (AR), a reliable clustering algorithm, has been successfully used in many fields [20–22]. The EMD method is particularly powerful for extracting the components of the basic mode from nonlinear or non-stationary time series [23], *i.e.*, the original complex time series can be transferred into a series of single and apparent components. However, this method cannot deal well with the signal decomposition effects while the gradient of the time series is fluctuating. Based on the empirical decomposition mode, reference [24] proposes the differential empirical mode decomposition (DEMD) to improve the fluctuating changes problem of the original EMD method. The derived signal is obtained by several derivations of the original signal, and the fluctuating gradient is thus eliminated, so that the signal can satisfy the conditions of EMD. The new signal is then integrated into EMD to obtain each intrinsic mode function (IMF) order and the residual amount of the original signal. The differential EMD method is employed to decompose the electric load into several detailed parts with higher frequency IMF and an approximate part with lower frequencies. This can effectively reduce the unnecessary interactions among singular values and can improve the performance when a single kernel function is used in forecasting. Therefore, it is beneficial to apply a suitable kernel function to conduct time series forecasting [25]. Since 1995, many attempts have been made to improve the performance of the PSO [26–31]. Sun *et al.* [32,33] introduced quantum theory into PSO and proposed a quantum-behaved PSO (QPSO) algorithm, which is a global search algorithm to theoretically guarantee finding good optimal solutions in the search space. Compared with PSO, the iterative equation of QPSO needs no velocity vectors for particles, has fewer parameters to adjust, and can be implemented more easily. The results of experiments on widely used benchmark functions indicate that the QPSO is a promising algorithm [32,33] that exhibits better performance than the standard PSO.

In this paper, we present a new hybrid model to achieve satisfactory forecasting accuracy. The principal idea is hybridizing DEMD with QPSO, SVR and AR, namely the DEMD-QPSO-SVR-AR model, to achieve better forecasting performance. The outline of the proposed DEMD-QPSO-SVR-AR model is as follows: (1) the raw data can be divided into two parts by DEMD technology, one is the higher frequency item, the other is the residuals; (2) the higher frequency item has less redundant information than the raw data and trend information, because that information is gone to the residuals, then, QPSO is applied to optimize the parameters of SVR (*i.e.*, the so-called QPSO-SVR model), so the QPSO-SVR model is used to forecast the higher frequency, the accuracy is higher than the original SVR model, particularly around the peak value; (3) fortunately, the residuals is monotonous and stationary, so the AR model is appropriate for forecasting the residuals; (4) the forecasting results are obtained from Steps (2) and (3). The proposed DEMD-QPSO-SVR-AR model has the capability of smoothing and reducing the noise (inherited from DEMD), the capability of filtering dataset and improving forecasting performance (inherited from SVR), and the capability of effectively forecasting the future tendencies (inherited from AR). The forecast outputs obtained by using the proposed hybrid method are described in the following sections.

To show the applicability and superiority of the proposed model, half-hourly electric load data (48 data points per day) from New South Wales (Australia) with two kind of sizes are used to compare the forecasting performances among the proposed model and other four alternative models, namely the PSO-BP model (BP neural network trained by the PSO algorithm), SVR model, PSO-SVR model (optimizing SVR parameters by the PSO algorithm), and the AFCM model (adaptive fuzzy combination model based on a self-organizing mapping and SVR). Secondly, another hourly electric load dataset (24 data points per day) from the New York Independent System Operator (NYISO, USA), also, with two kinds of sizes are used to further compare the forecasting performances of the proposed model with other three alternative models, namely the ARIMA model, BPNN model (artificial neural network trained by a back-propagation algorithm), and GA-ANN model (artificial neural network trained by a genetic algorithm). The experimental results indicate that this proposed DEMD-QPSO-SVR-AR model has the following advantages: (1) it simultaneously satisfies the need for high levels of accuracy and interpretability; (2) the proposed model can tolerate more redundant information than the SVR model, thus, it has more powerful generalization ability.

The rest of this paper is organized as follows: in Section 2, the DEMD-QPSO-SVR-AR forecasting model is introduced and the detailed illustrations of the model are also provided. In Section 3, the data description and the research design are illustrated. The numerical results and comparisons are shown in Section 4. The conclusions of this paper and the future research focuses are given in Section 5.

2. Support Vector Regression with Differential Empirical Mode Decomposition

2.1. Differential Empirical Mode Decomposition (DEMD)

The EMD method assumes that any signal consists of different simple intrinsic modes of oscillation. Each linear or non-linear mode will have the same number of extreme and zero-crossings. There is only one extreme between successive zero-crossings. In this way, each signal could be decomposed into a number of intrinsic mode functions (IMFs). With the definition, any signal $x(t)$ can be decomposed, and the corresponding flow chart is shown as Figure 1:

(1)   Identify all local extremes.
(2)   Repeat the procedure for the local minima to produce the lower envelope m_1.
(3)   The difference between the signal $x(t)$ and m_1 is the first component, h_1, as shown in Equation (1):

$$h_1 = x(t) - m_1 \tag{1}$$

In general, h_1 is unnecessary to satisfy the conditions of the IMF, because h_1 is not a standard IMF, and until the mean envelope approximates zero it should be determined k times. At this point, the data could be as shown in Equation (2):

$$h_{1k} = h_{1(k-1)} - m_{1k} \tag{2}$$

where h_{1k} is the datum after k siftings. $h_{1(k-1)}$ stands for the data after shifting $k - 1$ times. Standard deviation (SD) is defined by Equation (3):

$$SD = \sum_{k=1}^{T} \frac{\left| h_{1(k-1)}(t) - h_{1k}(t) \right|^2}{h_{1(k-1)}^2(t)} \in (0.2, 0.3) \tag{3}$$

where T is the length of the data.

(4)   When h_{1k} has met the basic conditions of SD, based on the condition of $c_1 = h_{1k}$, and a new series r_1 could be presented as Equation (4):

$$r_1 = x_1(t) - c_1 \tag{4}$$

(5)   Repeat previous steps 1 to 4 until the r_n cannot be decomposed into the IMF. The sequence r_n is called the remainder of the original data $x(t)$ as Equations (5) and (6):

$$r_1 = x_1(t) - c_1, r_2 = r_1 - c_2, ..., r_n = r_{n-1} - c_n \tag{5}$$

$$x_1(t) = \sum_{i=1}^{n} c_i + r_n \tag{6}$$

Finally, the differential EMD is proposed by Equation (7):

$$DEMD = x_n(t) - c_0(t) \tag{7}$$

where $x_n(t)$ refers to dependent variables.

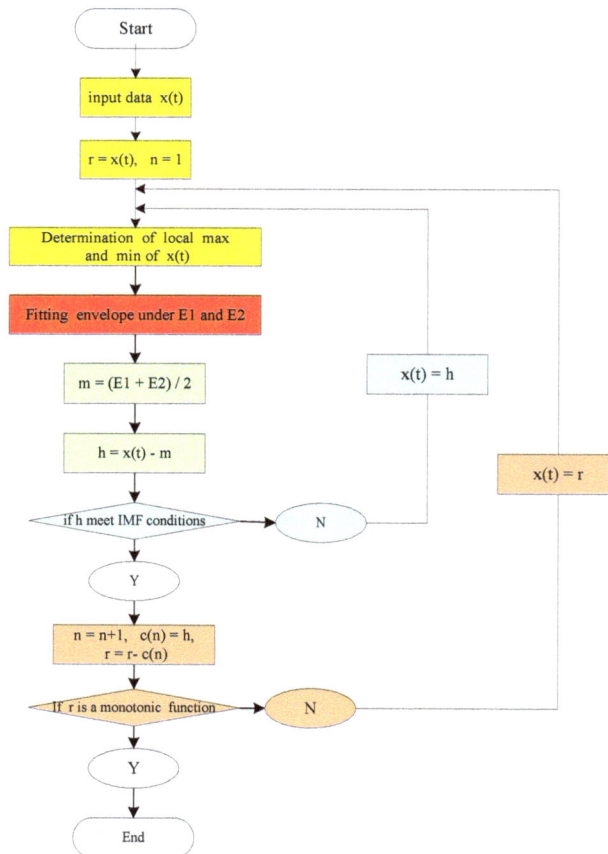

Figure 1. Differential EMD algorithm flowchart.

2.2. Support Vector Regression

The notion of an SVR model is briefly introduced. Given a data set with N elements $\{(X_i, y_i), i = 1, 2, ..., N\}$, where X_i is the i-th element in n-dimensional space, i.e., $X_i = [x_{1i}, ...x_{ni}] \in \Re^n$, and $y_i \in \Re$ is the actual value corresponding to X_i. A non-linear mapping $(\cdot) : \Re^n \rightarrow \Re^{n_h}$ is defined to map the training (input) data X_i into the so-called high dimensional feature space (which may have infinite dimensions), $\Re^{n_h}$. Then, in the high dimensional feature space, there theoretically exists a linear function, f, to formulate the non-linear relationship between input data and output data. Such a linear function, namely SVR function, is shown as Equation (8): align equations

$$f(X) = W^T \varphi(X) + b \tag{8}$$

where $f(X)$ denotes the forecasting values; the coefficients W ($W \in \Re^{n_h}$) and b ($b \in \Re$) are adjustable. As mentioned above, the SVM method aims at minimizing the empirical risk, shown as Equation (9):

$$R_{emp}(f) = \frac{1}{N} \sum_{i=1}^{N} \Theta_\varepsilon(y_i, W^T \phi(X_i) + b) \tag{9}$$

where $\Theta_\varepsilon(y_i, f(X))$ is the ε-insensitive loss function and defined as Equation (10):

$$\Theta_\varepsilon(y_i, f(X)) = \begin{cases} |f(X) - y| - \varepsilon & , \quad if \, |f(X) - y| \geqslant \varepsilon \\ 0 & , \quad otherwise \end{cases} \tag{10}$$

In addition, $\Theta_\varepsilon(y_i, f(X))$ is employed to find out an optimum hyper-plane on the high dimensional feature space to maximize the distance separating the training data into two subsets. Thus, the SVR focuses on finding the optimum hyperplane and minimizing the training error between the training data and the ε-insensitive loss function. Then, the SVR minimizes the overall errors, shown as Equation (11):

$$\underset{W, b, \xi^*, \xi}{Min} R_\varepsilon(W, \xi^*, \xi) = \frac{1}{2} W^T W + C \sum_{i=1}^{N} (\xi_i^* + \xi_i) \tag{11}$$

with the constraints:

$$\begin{aligned} y_i - W^T \phi(X_i) - b &\leqslant \varepsilon + \xi_i^* \\ -y_i + W^T \phi(X_i) + b &\leqslant \varepsilon + \xi_i \\ \xi_i^*, \xi_i &\geqslant 0 \\ i &= 1, 2, ..., N \end{aligned} \tag{12}$$

The first term of Equation (11), employing the concept of maximizing the distance of two separated training data, is used to regularize weight sizes to penalize large weights, and to maintain regression function flatness. The second term penalizes training errors of $f(x)$ and y by using the ε-insensitive loss function. C is the parameter to trade off these two terms. Training errors above ε are denoted as ξ_i^*, whereas training errors below $-\varepsilon$ are denoted as ξ_i.

After the quadratic optimization problem with inequality constraints is solved, the parameter vector w in Equation (8) is obtained as Equation (13):

$$W = \sum_{i=1}^{N} (\beta_i^* - \beta_i) \phi(X_i) \tag{13}$$

where β_i^*, ξ_i are obtained by solving a quadratic program and are the Lagrangian multipliers. Finally, the SVR regression function is obtained as Equation (14) in the dual space:

$$f(X) = \sum_{i=1}^{N} (\beta_i^* - \beta_i) K(X_i, X) + b \tag{14}$$

where $K(X_i,X)$ is called the kernel function, and the value of the kernel equals the inner product of two vectors, X_i and X_j, in the feature space $\varphi(X_i)$ and $\varphi(X_j)$, respectively; that is, $K(X_i,X_j) = \varphi(X_i)\varphi(X_j)$. Any function that meets Mercer's condition [34] can be used as the kernel function.

There are several types of kernel function. The most used kernel functions are the Gaussian radial basis functions (RBF) with a width of σ: $K(X)_i, X_j) = \exp\left(-0.5\|X_i - X_j\|^2/\sigma^2\right)$ and the polynomial kernel with an order of d and constants a_1 and a_2: $K(X_i,X_j) = (a_1X_i + a_2X_j)^d$. However, the Gaussian RBF kernel is not only easy to implement, but also capable of non-linearly mapping the training data into an infinite dimensional space, thus, it is suitable to deal with non-linear relationship problems. Therefore, the Gaussian RBF kernel function is specified in this study.

2.3. Particle Swarm Optimization Algorithm

PSO is a heuristic global optimization algorithm broadly applied in optimization problems. PSO is developed on a very simple theoretical framework that is easily implemented with only primitive mathematical operators [26]. In PSO, a group of particles is composed of m particles in D dimension space where the position of the particle i is $X_i = (x_{i1}, x_{i2}, \ldots, x_{iD})$ and the speed is $V_i = (v_{i1}, v_{i2}, \ldots, v_{iD})$. The speed and position of each particle are changed in accordance with the following equations, Equations (15) and (16):

$$v_{id}^{j+1} = wv_{id}^j + c_1r_1(p_{id}^j - x_{id}^j) + c_2r_2(p_{gd}^j - x_{id}^j) \tag{15}$$

$$x_{id}^{j+1} = x_{id}^j + v_{id}^{j+1} \tag{16}$$

where $i = 1, 2, \ldots, m$; $d = 1,2,\ldots, D$; m is the particle size; p_{id}^j is the dth dimension component of the *pbest* that is the individual optimal location of the particle i in the jth iteration; p_{gd}^j is the dth dimension component of the *gbest* that is the optimal position of all particles in the jth iteration; w is the inertia weight coefficient; c_1 and c_2 are learning factors; r_1 and r_2 are random numbers in the range $[0,1]$.

The inertia weight w, which balances the global and local exploitation abilities of the swarm, is critical for the performance of PSO. A large inertia weight facilitates exploration but slows down particle convergence. Conversely, a small inertia weight facilitates fast particle convergence it sometimes leads to the local optimal. The most popular algorithm for controlling inertia weight is linearly decreasing inertia weight PSO [31]. The strategy of linearly decreasing inertia weight is widely used to improve the performance of PSO, but this approach has a number of drawbacks [27]. Several adaptive algorithms for tuning inertia weight have been presented [27–30]. In the present work, we propose the method of nonlinearly decreasing inertia weight to tune the value of w for further performance improvement as Equation (17):

$$w = w_{\max} - \frac{(w_{\max} - w_{\min}) \times (t-1)^2}{(t_{\max}-1)^2} \tag{17}$$

where $w_{\max}$ and $w_{\min}$ are the maximum and minimum values of w, respectively; t is the current iteration number; and $t_{\max}$ is the maximum iteration number.

2.4. Quantum-Behaved Particle Swarm Optimization

The main disadvantage of the PSO algorithm is that global convergence is not guaranteed [35]. To address this problem, Sun *et al.* [32,33], inspired by the trajectory analysis of PSO and quantum mechanics, developed and proposed the QPSO algorithm. Particles move according to the following iterative equations, Equations (18) to (21), and the flow chart is shown as Figure 2.

$$x_{ij}(t+1) = p_{ij}(t) + \alpha \left| mbest_j(t) - x_{ij}(t) \right| \times \ln(1/u) \qquad if\ k \geqslant 0.5 \tag{18}$$

$$x_{ij}(t+1) = p_{ij}(t) - \alpha \left| mbest_j(t) - x_{ij}(t) \right| \times \ln(1/u) \qquad if\ k \leqslant 0.5 \tag{19}$$

$$mbest_j(t) = \frac{1}{M}\sum_{i=1}^{M} pbest_{ij}(t) \tag{20}$$

$$p_{ij}(t) = \phi_{ij}(t)pbest_{ij}(t) + (1 - \phi_{ij}(t))gbest_j(t) \tag{21}$$

where *mbest* is the mean best position defined as the mean of all the *pbest* positions of the population; k, u and φu are random numbers generated using a uniform probability distribution in the range [0,1]. The parameter α is called the contraction expansion coefficient, which is the only parameter in the QPSO algorithm that can be tuned to control the convergence speed of particles. In general, this parameter can be controlled by two methods: (1) fixing; or (2) varying the value of α during the search of the algorithm. In [36], setting α to a number in the range (0.5, 0.8) generates satisfied results for most benchmark functions. However, fixing the value of α is sensitive to population size and the maximum number of iterations. This problem can be overcome by using a time-varying CE coefficient. The literatures on QPSO suggest that decreasing the value of α linearly from α_1 to α_0 ($\alpha_0 < \alpha_1$) in the course of the search process makes the QPSO algorithm perform efficiently [36,37].

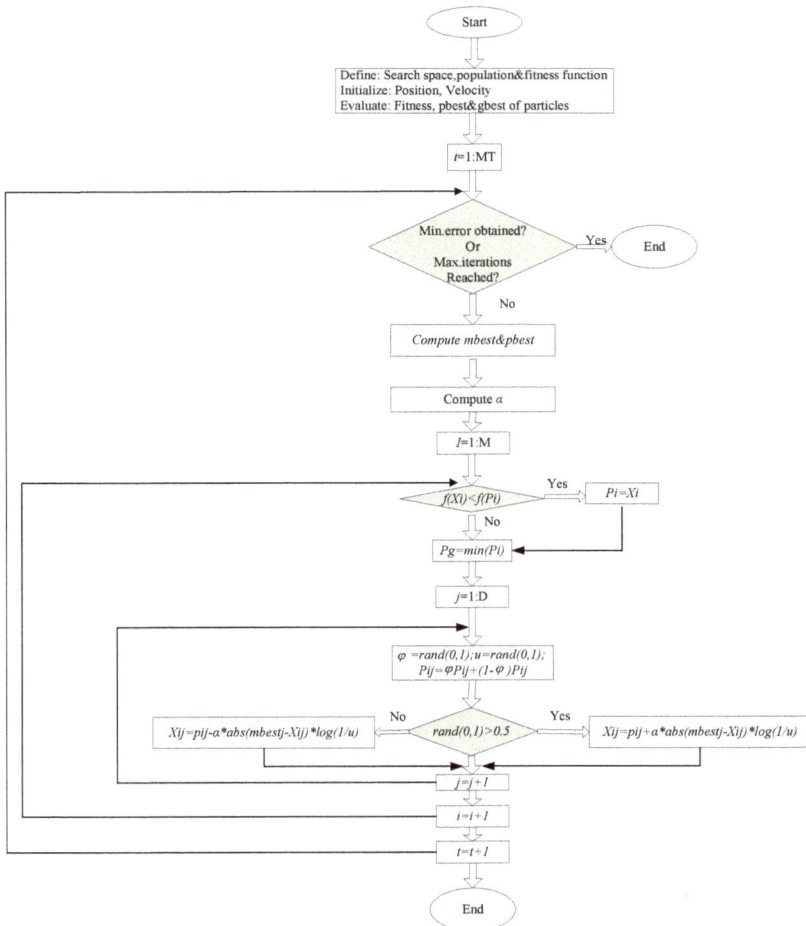

Figure 2. QPSO algorithm flowchart.

In this paper, the value of α is computed as Equation (22):

$$\alpha = \alpha_1 - \frac{(t-1) \times (\alpha_1 - \alpha_0)}{T-1} \tag{22}$$

where α_1 and α_0 are the final and initial values of α, respectively; t is the current iteration number; and T is the maximum iteration number. Previous studies on QPSO [36,37] recommend that α be linearly decreased from 1.0 to 0.5 for the algorithm to attain a generally good performance.

QPSO has already been implemented with excellent results [32] in various standard optimization problems. Moreover, the QPSO algorithm has been proven more effective than traditional algorithms in most cases [38–42]. In the current work, QPSO algorithm is utilized in SVR parameter optimization for forecasting the high frequency data, and its performance is compared with that of the classical PSO algorithm [43,44].

2.5. AR Model

Equation (23) expresses a p-step autoregressive model, referring as $AR(p)$ model [45]. Stationary time series $\{X_t\}$ that meet the model $AR(p)$ is called the $AR(p)$ sequence. That $a = (a_1, a_2, \ldots, a_p)^\mathrm{T}$ is named as the regression coefficients of the $AR(p)$ model:

$$X_t = \sum_{j=1}^{p} a_j X_{t-j} + \varepsilon_t \tag{23}$$
$$t \in Z$$

2.6. The Full Procedure of DEMD-QPSO-SVR-AR Model

The full procedure of the proposed DEMD-QPSO-SVR-AR model is briefed as follows and is illustrated in Figure 3.

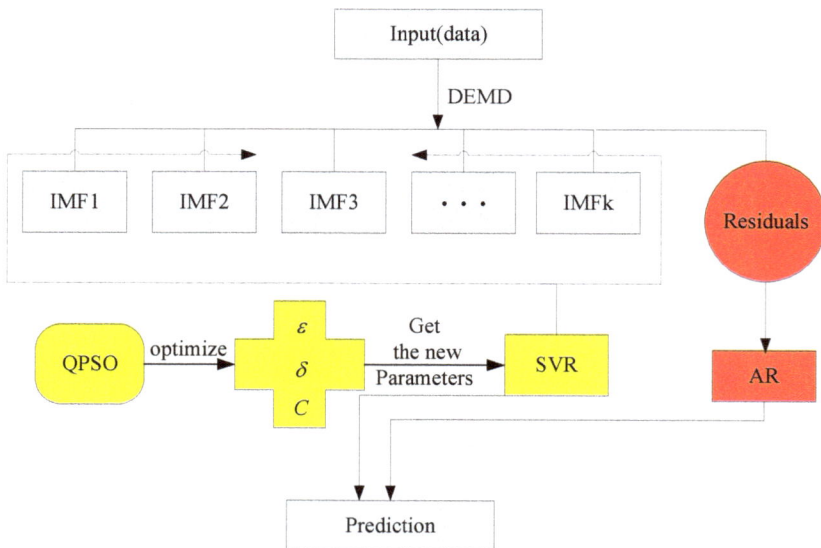

Figure 3. The full flowchart of the DEMD-QPSO-SVR-AR model flowchart.

Step 1: Decompose the input data by DEMD. Each electric load data (input data) could be decomposed into a number of intrinsic mode functions (IMFs), *i.e.*, two parts, one is the higher frequency item, the other one is the residuals. Please refer Section 2.1 and Figure 1 to learn more about the DEMD process details.

Step 2: QPSO-SVR modeling. The SVR model is employed to forecast the high frequency item, thus, to look for most suitable parameters by QPSO, different sizes of fed-in/fed-out subsets will be set in this stage. Please refer Section 2.2 to learn in more details the SVR process. The QPSO algorithm is utilized in SVR parameter optimization for forecasting the higher frequency data, please refer Sections 2.3 and 2.4 and Figure 2 to learn more about the details of the QPSO process.

Step 3: AR modeling. The residuals item is forecasted by the AR model due to its monotonous and stationary nature. Please refer Section 2.5 to learn in more detail the processes of AR modeling. Similarly, while the new parameters have smaller MAPE values or maximum iteration is reached, the new three parameters and the corresponding objective value are the solution at this stage.

Step 4: DEMD-QPSO-SVR-AR forecasting. After receiving the forecasting values of the high frequency item and the residuals item from SVR model and AR model, respectively, the final forecasting results would be eventually obtained from the high frequency item and the residuals.

3. Numerical Examples

To illustrate the superiority of the proposed model, we use two datasets from different electricity markets, that is, the New South Wales (NSW) market in Australia (denoted as Case 1) and the New York Independent System Operator (NYISO) in the USA (Case 2). In addition, for each case, we all use two sample sizes, called small sample and large sample, respectively.

3.1. The Experimental Results of Case 1

For Case 1, firstly, electric load data obtained from 2 to 7 May 2007 is used as the training data set in the modeling process, and the testing data set is from 8 May 2007. The electric load data used are all based on a half-hourly basis (*i.e.*, 48 data points per day). The dataset containing only 7 days is called the small size sample in this paper.

Secondly, for large training sets, it should avoid overtraining during the SVR modeling process. Therefore, the second data size has 23 days (1104 data points from 2 to 24 May 2007) by employing all of the training samples as training set, *i.e.*, from 2 to 17 May 2007, and the testing data set is from 18 to 24 May 2007. This example is called the large sample size data in this paper.

(i) Results after DEMD in Case 1

As mentioned in the authors' previous paper [25], the results of the decomposition process by DEMD, can be divided into the higher frequency item (Data-I) and the residuals term (Data-II). The trend of the higher frequency item is the same as that of the original data, and the structure is more regular and stable. Thus, Data-I and Data-II both have good regression effects by the QPSO-SVR and AR, respectively.

(ii) Forecasting Using QPSO-SVR for Data-I (The Higher Frequency Item in Case 1)

After employing DEMD to reduce the non-stationarity of the data set in Case 1, QPSO with SVR can be successfully applied to reduce the performance volatility of SVR with different parameters, to perform the parameter determination in SVR modeling process.

The higher frequency item is simultaneously employed for QPSO-SVR modeling, and the better performances of the training and testing (forecasting) sets are shown in Figure 4a,b, respectively. This implies that the decomposition and optimization by QPSO is helpful to improve the forecasting accuracy. The parameters of a QPSO-SVR model for Data-I are shown in Tables 1 and 2 in which the forecasting error for the higher frequency decomposed by the DEMD and QPSO-SVR has been reduced.

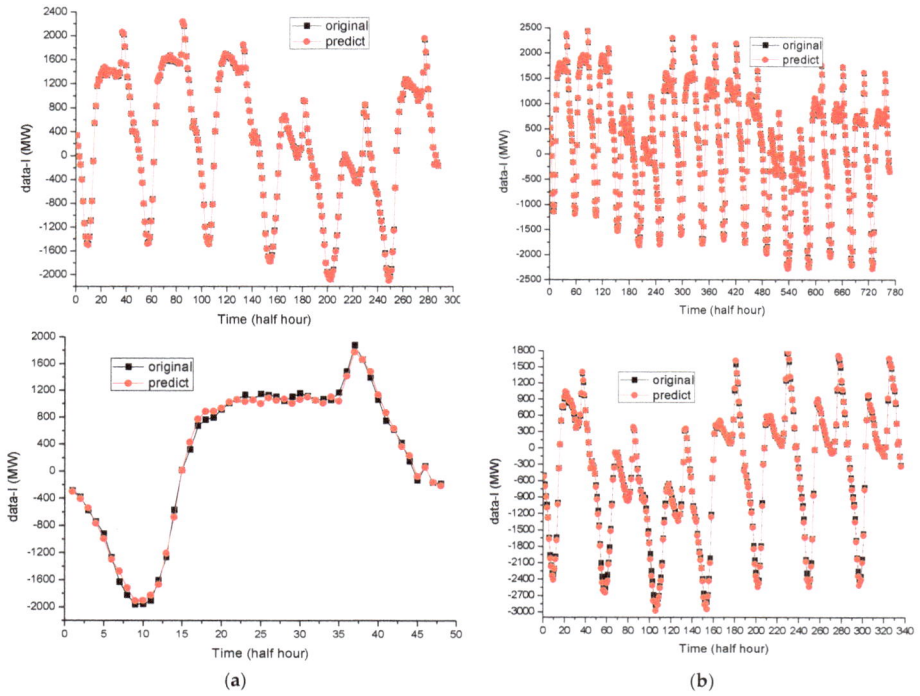

Figure 4. Comparison the forecasted electric load of train and test by the QPSO-SVR model for the data-I of sample data: (**a**) One-day ahead prediction of May 8, 2007 are performed by the model; (**b**) One-week ahead prediction from May 18, 2007 May 24, 2007 are performed by the model.

Table 1. The QPSO's parameters for SVR in Data-I.

N	C_{min}	C_{max}	σ_{min}	σ_{max}	*itmax*
30	0	200	0	200	300

Table 2. The SVR's parameters for Data-I.

Sample Size	m	σ	C	ε	**Testing MAPE**
The samall sample data	20	0.12	88	0.0027	9.13
The large sample data	20	0.19	107	0.0011	4.1

where N is number of particles, C_{min} is the minimum of C, C_{max} is the maximum of C, σ_{min} is the minimum of σ, σ_{max} is the maximum of σ, *itmax* is maximum iteration number.

(iii) Forecasting Using AR for Data-II (The Residuals in Case 1)

As mentioned in the authors' previous paper [25], the residuals are linear locally and stable, so use of the AR technique to predict Data-II is feasible. Based on the geometric decay of the correlation analysis for Data-II (the residuals), it can be denoted as the AR(4) model. The associated parameters of the AR(4) model for Data-II are indicated in Table 3. The errors almost approach the level of 10^{-5} both for the small or large amounts of data, *i.e.*, the forecasting error for Data-II by DEMD has been significantly reduced. This shows the superiority of the AR model.

Table 3. Summary of results of the AR forecasting model for data-II.

Residuals	MAE	Equation
The small sample size	9.7732×10^{-5}	$x_n = 5523.827 + 1.23x_{n-1} + 0.5726x_{n-2} + 0.0031x_{n-3} - 0.70465x_{n-4}$
The large sample size	7.5923×10^{-5}	$x_n = 5524.9 + 1.0152x_{n-1} + 0.3628x_{n-2} + 0.0019x_{n-3} - 0.6752x_{n-4}$

3.2. The Experimental Results of Case 2

For Case 2, electric load data obtained from 1 to 12 January 2015 is used as the training data set in the modeling process, and the testing data set is from 13 to 14 January 2015. These employed electric load data are all based on an hour basis (*i.e.*, 24 data points per day). The dataset contains only 14 days so it is also called the small sample in this paper.

Secondly, for large training sets, the second dataset size is 46 days (1104 data points from 1 January to 15 February 2015) by employing all of the training samples as training set, *i.e.*, from 1 January to 1 February 2015, and the testing dataset is from 2 to 15 February 2015. This example is also called the large size sample data in this paper.

(i) Results after DEMD in Case 2

As mentioned in the authors' previous paper [25], similarly, the data results of the decomposition process by DEMD can be divided into the higher frequency item (Data-I) and the residuals term (Data-II). The trend of the higher frequency item is also the same as that of the original data, and the structure is also regular and stable. Thus, Data-I and Data-II both have good regression effects by the QPSO-SVR and AR, respectively.

(ii) Forecasting Using QPSO-SVR for Data-I (The Higher Frequency Item in Case 2)

After employing DEMD to reduce the non-stationarity of the data set in Case 2, to further resolve these complex nonlinear, chaotic problems for both small sample and large sample data, the QPSO with SVR can be successfully applied to reduce the performance volatility of SVR with different parameters to perform the parameter determination in the SVR modeling process, to improve the forecasting accuracy. The higher frequency item is simultaneously employed for QPSO-SVR modeling, and the better performances of the training and testing (forecasting) sets are shown in Figure 5a,b, respectively. This implies that the decomposition and optimization by QPSO is helpful to improve the forecasting accuracy. The parameters of a QPSO-SVR model for Data-I are shown in Tables 1 and 4 in which the forecasting error for the higher frequency decomposed by the DEMD and QPSO-SVR has been reduced.

Table 4. The SVR's parameters for data-I in Case 2.

Sample Size	m	σ	C	ε	Testing MAPE
The small data	24	0.10	102	0.0029	7.19
The large data	24	0.19	113	0.0011	4.62

(iii) Forecasting Using AR for Data-II (The Residuals in Case 2)

As mentioned in the authors' previous paper [25], the residuals are linear locally and stable, so the AR technique is feasible to predict Data-II. Based on the geometric decay of the correlation analysis for Data-II (the residuals), that can also be denoted as the AR(4) model, the associated parameters of the AR(4) model for Data-II are indicated in Table 5. The errors almost approach a level of 10^{-5} both for the small or large amount of data, *i.e.*, the forecasting error for Data-II by DEMD has significantly reduced. This shows the superiority of the AR model.

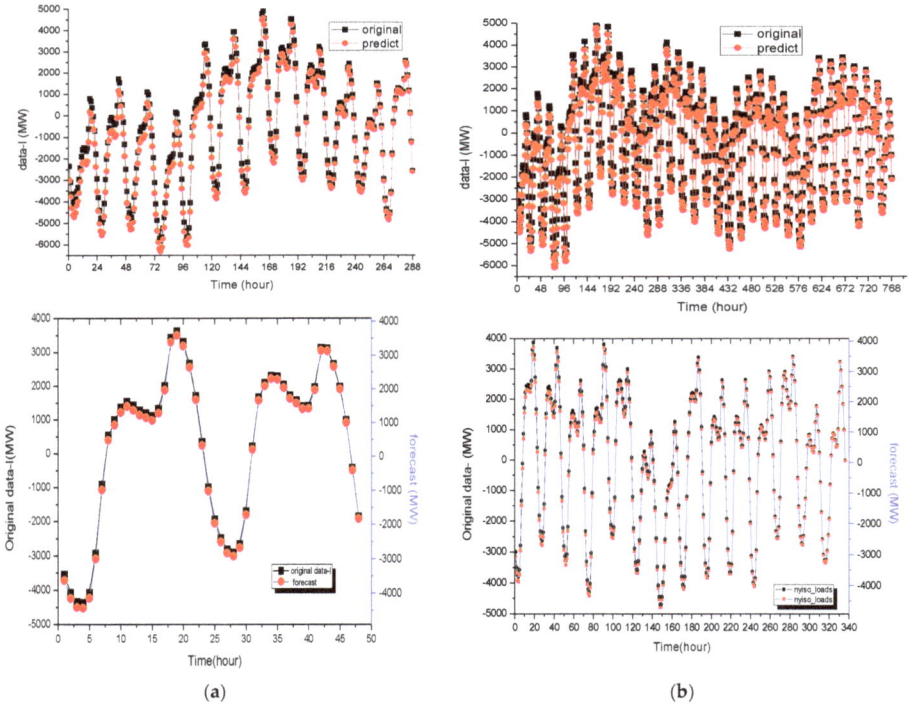

Figure 5. Comparison the forecasted electric load of training and test data by the QPSO-SVR model for the data-I of sample data in Case 2: (**a**) One-day ahead prediction from 13 to 14 January 2015 are performed by the model; (**b**) One-week ahead prediction from 2 to 15 February 2015 are performed by the model.

Table 5. Summary of results of the AR forecasting model for Data-II in Case 2.

Residuals	MAE	Equation
The small sample size	9.138×10^{-5}	$x_n = 5521.7 + 1.13x_{n-1} + 0.5676x_{n-2} + 0.021x_{n-3} - 0.845x_{n-4}$
The large sample size	6.02×10^{-5}	$x_n = 5522.7 + 0.9152x_{n-1} + 0.3978x_{n-2} + 0.0049x_{n-3} - 0.52x_{n-4}$

where x_n is the *n*-th electric load residual, x_{n-1} is the $(n-1)$th electric load residual similarly, *etc.*

4. Results and Analysis

This section illustrates the performance of the proposed DEMD-QPSO-SVR-AR model in terms of forecasting accuracy and interpretability. Taking into account the superiority of an SVR model for small sample size and superiority comparisons, a real case analysis with small sample size is used in the first case. The next case with 1104 data points is devoted to illustrate the relationships between two sample sizes (large size and small size) and accurate levels in forecasting.

4.1. Setting Parameters for the Proposed Forecasting Models

As indicated by Taylor [46], and according to the same conditions of the comparison with Che *et al.* [47], the settings of several parameters in the proposed forecasting models are illustrated as follows. For the PSO-BP model, 90% of the collected samples is used to train the model, and the rest (10%) is employed to test the performance. In the PSO-BP model, these used parameters are set as follows: (i) for the BP neural network, the input layer dimension (*indim*) is set as 2; the dimension of the hidden layer (*hiddennum*) is set as 3; the dimension of the output layer (*outdim*) is set as 1; (ii) for the PSO algorithm, the maximum iteration number (*itmax*) is set as 300; the number of the searching particles, N, is set as 40; the length of each particle, D, is set as 3; weight c_1 and c_2 are set as 2.

The PSO-SVR model not only has its embedded constraints and limitations from the original SVR model, but also has huge iteration steps as a result of the requirements of the PSO algorithm. Therefore, it would be time consuming to train the PSO-SVR model while the total training set is used. For this consideration, the total training set is divided into two sub-sets, namely training subset and evaluation subset. In the PSO algorithm, the parameters used are set as follows: for the small sample, the maximum iteration number (*itmax*) is set as 50; the number of the searching particles, N, is set as 20; the length of each particle, D, is set as 3; weight c_1 and c_2 are set as 2; for the large sample, the maximum iteration number (*itmax*) is set as 20; the number of the searching particles, N, is set as 5; the length of each particle, D, is also set as 3; weight c_1 and c_2 are also set as 2.

Regarding Case 2, to be based on the same comparison conditions used in Fan *et al.* [25], the newest electric load data from NYISO is also employed for modeling, five alternative forecasting models (including the ARIMA, BPNN, GA-ANN, EMD-SVR-AR, and DEMD-SVR-AR models) are used for comparison with the proposed model. Some parameter settings of the employed forecasting models are set the same as in [25], and are briefly as follows: for the BPNN model, the node numbers of its structure are different for small sample size and large sample size; for the former one, the input layer dimension is 240, the hidden layer dimension is 12, and the output layer dimension is 48; and these values are 480, 12, 336, respectively, for the latter one. The parameters of GA-ANN model used in this case are as follows: generation number is set as 5, population size is set as 100, bit numbers are set as 50, mutation rate is set as 0.8, crossover rate is 0.05.

4.2. Evaluation Indices for Forecasting Performances

For evaluating the forecasting performances, three famous forecasting accurate level indices, RMSE (root mean square error), MAE (mean absolute error), and MAPE (mean absolute percentage error), as shown in Equations (24) to (26), are employed:

$$RMSE = \sqrt{\frac{\sum_{i=1}^{n} (P_i - A_i)^2}{n}} \tag{24}$$

$$MAE = \frac{\sum_{i=1}^{n} |P_i - A_i|}{n} \tag{25}$$

$$MAPE = \frac{\sum_{i=1}^{n} \left| \frac{P_i - A_i}{A_i} \right|}{n} * 100 \tag{26}$$

where P_i and A_i are the *i*-th forecasting and actual values, respectively, and n is the total number of forecasts.

In addition, to verify the suitability of model selection, Akaike's Information Criterion (AIC), an index of measurement for the relative quality of models for a given set of data, and Bayesian Information Criterion (BIC), also known as the Schwartz criterion, which is a criterion for model selection among a finite set of models (the model with the lowest BIC is preferred), are both taken into account to enhance the robustness of the verification. These two indices are defined as Equations (27) and (28), respectively:

$$AIC = Log(SSE) + 2q \tag{27}$$

where *SSE* is the sum of squares for errors, *q* is the number of estimated parameters:

$$BIC = Log(SSE) + qLog(n) \tag{28}$$

where *q* is the number of estimated parameters and *n* is the sample size.

4.3. Empirical Results and Analysis

For the first experiment in Case 1, the forecasting results (the electric load on 8 May 2007) of the original SVR model, the PSO-SVR model and the proposed DEMD-QPSO-SVR-AR model are shown in Figure 6a. For Case 2, the forecasting results of the ARIMA model, the BPNN model, the GA-ANN model and the proposed DEMD-QPSO-SVR-AR model are shown in Figure 7a. Based on these two figures, the forecasting curve of the proposed DEMD-QPSO-SVR-AR model seems to achieve a better fit than other alternative models for the two cases in this experiment.

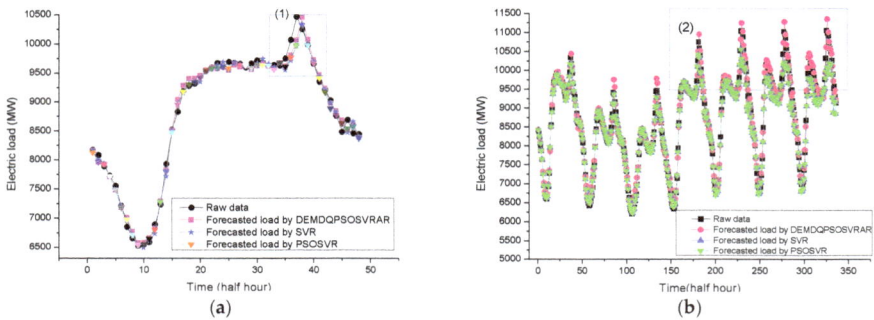

Figure 6. Comparison of the original data and the forecasted electric load by the DEMD-QPSO-SVR-AR Model, the SVR model and the PSO-SVR model for (**a**) the small sample size (One-day ahead prediction of 8 May 2007 are performed by the models); (**b**) the large sample size (One-week ahead prediction from 18 May 2007 to 24 May 2007 are performed by the models).

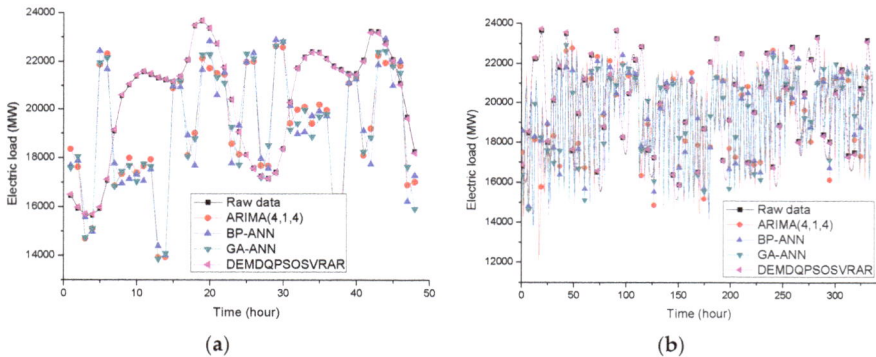

Figure 7. Comparison of the original data and the forecasted electric load by the DEMD-QPSO-SVR-AR Model, the ARIMA model, the BPNN model and the GA-ANN model for: (**a**) the small sample size (One-day ahead prediction from 13 to 14 January 2015 are performed by the models); (**b**) the large sample size (One-week ahead prediction from 2 to15 February 2015 are performed by the models).

The second experiments in Cases 1 and 2 show the large sample size data. The peak load values of the testing set are bigger than those of the training set. The detailed forecasting results in this experiment are illustrated in Figures 6b and 7b. It also shows that the results obtained from the proposed DEMD-QPSO-SVR-AR model seem to have smaller forecasting errors than other alternative models.

Notice that for any particular sharp points in Figures 6 and 7 after extracting the direction feature of the trend by DEMD technology, these sharp points fixed in their positions represent the higher frequency characteristics of the remaining term, therefore, quantizing the particles in PSO the algorithm is very effective for dealing with this kind of fixed point characteristics. In other words, the DEMD-QPSO-SVR-AR model has better generalization ability than other alternative comparison models in both cases. Particularly in Case 1, for example, the local details for sharp points in Figure 6a,b are enlarged and are shown in Figure 8a,b, respectively. It is clear that the forecasting curve of the proposed DEMD-QPSO-SVR-AR model (red solid dots and red curve) fits more precisely than other alternative models, *i.e.*, it is superior for capturing the data change trends, including any fluctuation tendency.

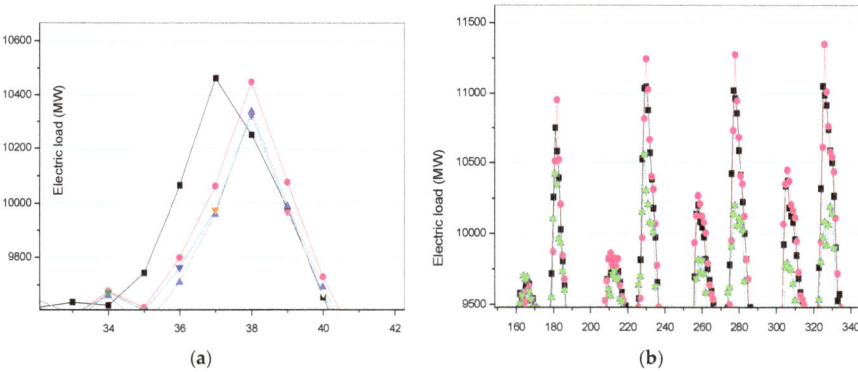

Figure 8. The local enlargement (peak) comparison of the DEMDQPSOSVRAR Model, the SVR model and the PSO-SVR model for (**a**) the small sample size *(1)*; (**b**) the large sample size *(2)*.

To better explain the superiority, the shape factor (*SF*), defined as Equation (29) and shown in Figure 9, is employed to illustrate the fitting effectiveness of the method, the *SF* value of the model closer to the one of the raw data, the fitness of the model is better than others. The results are shown in Table 6. It indicates that the data of *SF* from DEMD-QPSO-SVR-AR model is closer to the raw data than other models:

$$SF\,(\text{Shape Factor})\;=\;S_1(\text{Square of blue area})/S_2(\text{Square of red area}) \tag{29}$$

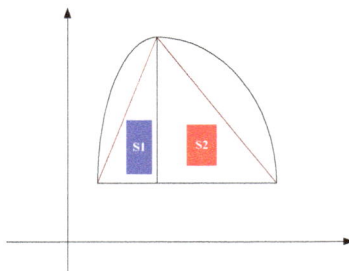

Figure 9. The definition of shape factor.

Table 6. The shape factor (*SF*) for local sharp points comparison.

Algorithms	Small Sample Size	Large Sample Size								
	SF	SF1	SF2	SF3	SF4	SF5	SF6	SF7	SF8	MSF *
Raw data	0.83	0.92	0.95	0.81	0.76	0.73	0.71	0.74	0.71	0.791
Original SVR	1.05	1.13	1.02	0.92	0.88	0.90	0.85	0.96	0.83	0.924
PSO-SVR	1.12	0.99	0.82	0.89	0.85	0.87	0.64	0.72	0.79	0.821
DEMD-QPSO-SVR-AR	0.90	0.94	0.93	0.81	0.75	0.69	0.74	0.72	0.70	0.785

* *MSF* = mean(*SF*).

The forecasting results in Cases 1 and 2 are summarized in Tables 7 and 8 respectively. The proposed DEMD-QPSO-SVR-AR model is compared with alternative models. It is indicated that our hybrid model outperforms all other alternatives in terms of all the evaluation criteria. One of the general observations is that the proposed model tends to fit closer to the actual value with a smaller forecasting error. This is ascribed to the fact that a well combined DEMD and QPSO can effectively capture the exact shape characteristics, which are difficult to illustrate by many other methods while data often has intertwined effects among the chaos, noise, and other unstable factors. Therefore, the unstable impact is well solved by DEMD, especially for those border points, and then, QPSO can accurately illustrate the chaotic rules, *i.e.*, achieve more satisfactory parameter solutions for an SVR model.

Table 7. Summary of results of the forecasting models in Case 1.

Algorithm	MAPE	RMSE	MAE	AIC	BIC	Running Time (s)	MSF (Data)
For the first experiment (small sample size)							
Original SVR [47]	11.6955	145.865	10.9181	112.3	111.9	180.4	0.972 (0.749)
PSO-SVR [47]	11.4189	145.685	10.6739	120.7	125.8	165.2	0.904 (0.749)
PSO-BP [47]	10.9094	142.261	10.1429	110.5	116.0	159.9	0.897 (0.749)
AFCM [23]	9.9524	125.323	9.2588	82.6	85.5	75.3	0.761 (0.749)
DEMD-QPSO-SVR-AR	9.1325	122.368	9.2201	80.9	83.1	100.7	0.756 (0.749)
For the second experiment (large sample size)							
Original SVR [47]	12.8765	181.617	12.0528	167.7	180.9	116.8	1.062 (0.830)
PSO-SVR [47]	13.503	271.429	13.0739	215.8	220.3	192.7	0.994 (0.830)
PSO-BP [47]	12.2384	175.235	11.3555	150.4	157.2	163.1	0.925 (0.830)
AFCM [23]	11.1019	158.754	10.4385	142.1	146.7	160.4	0.838 (0.830)
DEMD-QPSO-SVR-AR	4.1499	140.105	9.6258	129.0	128.1	169.0	0.826 (0.830)

In view of the model effectiveness and efficiency on the whole, we can conclude that the proposed model is quite competitive against other compared models, such as the ARIMA, BPNN, GA-ANN, PSO-BP, SVR, PSO-SVR, and AFCM models. In other words, the hybrid model leads to better accuracy and statistical interpretation.

In particular, as shown in Figure 8, our method shows higher accuracy and good flexibility in peak or inflection points, because the little redundant information could be used by statistical learning or regression models, and the level of optimization would increase. This also ensures that it could achieve more significant forecasting results due to the closer *SF* values to the raw data set. For closer insight, this can be viewed as the fact that the shape factor reflects how the electric load demand mechanism ia affected by multiple factors, *i.e.*, the shape factor reflects the change tendency in terms of ups or downs, thus, closer SF value to the raw data set can capture more precise trend changes than others, and this method no doubt can reveal the regularities for any point status.

Table 8. Summary of results of the forecasting models in Case 2.

Algorithm	MAPE	RMSE	MAE	AIC	BIC	MSF (Data)
For the first experiment (small sample size)						
ARIMA(4,1,4) [25]	45.33	320.45	25.72	278.4	285.1	1.132 (0.864)
BPNN [25]	31.76	219.43	21.69	198.5	200.2	0.955 (0.864)
GA-ANN [25]	23.89	220.96	23.55	199.3	202.3	0.947 (0.864)
EMD-SVR-AR [25]	14.31	158.11	17.44	140.7	141.9	0.873 (0.864)
DEMD-SVR-AR [25]	8.19	140.16	12.79	128.3	130.1	0.865 (0.864)
DEMD-QPSO-SVR-AR	7.48	138.89	14.44	125.4	126.7	0.859 (0.864)
For the second experiment (large sample size)						
ARIMA(4,1,4) [25]	60. 65	733.22	54.05	551.4	579.4	1.091 (0.875)
BPNN [25]	42.5	479.48	50.39	334.6	342.3	0.976 (0.875)
GA-ANN [25]	33.12	450.63	44.35	321.7	323.6	0.953 (0.875)
EMD-SVR-AR [25]	11.29	289.21	20.76	239.0	238.2	0.890 (0.875)
DEMD-SVR-AR [25]	5.37	160.58	15.82	141.4	142.2	0.884 (0.875)
DEMD-QPSO-SVR-AR	4.62	153.22	16.30	132.5	133.8	0.879 (0.875)

Several findings deserved to be noted. Firstly, based on the forecasting performance comparisons among these models, the proposed model outperforms other alternative models. Secondly, the proposed model has better generalization ability for different input patterns as shown in the second experiment. Thirdly, from the comparison between the different sample sizes of these two experiments, we conclude that the hybrid model can tolerate more redundant information and construct the model for the larger sample size data set. Fourthly, based on the calculation and comparison of *SF* in Table 6, the proposed model also receives closer *SF* values to the raw data than other alternative models. Finally, since the proposed model generates good results with good accuracy and interpretability, it is robust and effective, as shown in Tables 7 and 8 comparing the other models, namely the original SVR, PSO-SVR, PSO-BP and AFCM models. Overall, the proposed model provides a very powerful tool that is easy to implement for electric load forecasting.

Eventually, the most important issue is to verify the significance of the accuracy improvement of the proposed model. The forecasting accuracy comparisons in both cases among original SVR, PSO-SVR, PSO-BP, AFCM, ARIMA, BPNN, and GA-ANN models are conducted by a statistical test, namely a Wilcoxon signed-rank test, at the 0.025 and 0.05 significant levels in one-tail-tests. The Wilcoxon signed-rank test is a non-parametric statistical hypothesis test used when comparing two related samples, matched samples, or repeated measurements on a single sample to assess whether their population mean ranks differ (*i.e.*, it is a paired difference test). It can be used as an alternative to the paired Student's t-test, t-test for matched pairs, or the *t*-test for dependent samples when the population cannot be assumed to be normally distributed [48]. The test results are shown in Tables 9 and 10. Clearly, the outstanding forecasting results achieved by the proposed model is only significantly superior to other alternative models at a significance level of 0.05. This also implies that there are still lots of improvement efforts that can be made for hybrid quantum-behavior evolutionary SVR-based models.

Table 9. Wilcoxon signed-rank test in Case 1.

Compared Models	Wilcoxon Signed-Rank Test	
	$\alpha = 0.025$; W = 4	$\alpha = 0.05$; W = 6
DEMDQPSOSVRAR *vs.* original SVR	8	3 [a]
DEMDQPSOSVRAR *vs.* PSO-SVR	6	2 [a]
DEMDQPSOSVRAR *vs.* PSO-BP	6	2 [a]
DEMDQPSOSVRAR *vs.* AFCM	6	2 [a]

[a] denotes that the DEMD-QPSO-SVR-AR model significantly outperforms other alternative models.

Table 10. Wilcoxon signed-rank test in Case 2.

Compared Models	Wilcoxon Signed-Rank Test	
	$\alpha = 0.025$; W = 4	$\alpha = 0.05$; W = 6
DEMDQPSOSVRAR *vs.* ARIMA	6	2 [a]
DEMDQPSOSVRAR *vs.* BPNN	6	2 [a]
DEMDQPSOSVRAR *vs.* GA-ANN	6	2 [a]
DEMDQPSOSVRAR *vs.* EMD-SVR-AR	6	2 [a]
DEMDQPSOSVRAR *vs.* DEMD-SVR-AR	6	2 [a]

[a] denotes that the DEMD-QPSO-SVR-AR model significantly outperforms other alternative models.

5. Conclusions

This paper presents an SVR model hybridized with the differential empirical mode decomposition (DEMD) method and quantum particle swarm optimization algorithm (QPSO) for electric load forecasting. The experimental results indicate that the proposed model is significantly superior to the original SVR, PSO-SVR, PSO-BP, AFCM, ARIMA, BPNN, and GA-ANN models. To improve the forecasting performance (accuracy level), quantum theory is hybridized with PSO (namely the QPSO) into an SVR model to determine its suitable parameter values. Furthermore, the DEMD is employed to simultaneously consider the accuracy and comprehensibility of the forecast results. Eventually, a hybrid model (namely DEMD-QPSO-SVR-AR model) has been proposed and its electric load forecasting superiority has also been compared with other alternative models. It is also demonstrated that a well combined DEMD and QPSO can effectively capture the exact shape characteristics, which are difficult to illustrate by many other methods while data often has intertwined effects among the chaos, noise, and other unstable factors. Hence, the instability impact can be well solved by DEMD, especially for those border points, and then, QPSO can accurately illustrate the chaotic rules, thus achieving more satisfactory parameter solutions than an SVR model.

Acknowledgments: Guo-Feng Fan thanks the research grants sponsored by National Natural Science Foundation of China (under Contract No. 51064015), Startup Foundation for Doctors (PXY-BSQD-2014001), and Educational Commission of Henan Province of China (15A530010); Li-Ling Peng thanks the research grant sponsored by Youth Foundation of Ping Ding Shan University (PXY-QNJJ-2014008); and Professor Wei-Chiang Hong and Min-Liang Huang thank the research grant sponsored by Ministry of Science & Technology, Taiwan (MOST 104-2410-H-161-002).

Author Contributions: Li-Ling Peng and Wei-Chiang Hong conceived and designed the experiments; Guo-Feng Fan and Min-Liang Huang performed the experiments; Li-Ling Peng and Wei-Chiang Hong analyzed the data; Li-Ling Peng wrote the paper.

Conflicts of Interest: The authors declare no conflict of interest.

References

1. Bernard, J.T.; Bolduc, D.; Yameogo, N.D.; Rahman, S. A pseudo-panel data model of household electricity demand. *Resour. Energy Econ.* **2010**, *33*, 315–325. [CrossRef]
2. Bianco, V.; Manca, O.; Nardini, S. Electricity consumption forecasting in Italy using linear regression models. *Energy* **2009**, *34*, 1413–1421. [CrossRef]

3. Zhou, P.; Ang, B.W.; Poh, K.L. A trigonometric grey prediction approach to forecasting electricity demand. *Energy* **2006**, *31*, 2839–2847. [CrossRef]
4. Afshar, K.; Bigdeli, N. Data analysis and short term load forecasting in Iran electricity market using singular spectral analysis (SSA). *Energy* **2011**, *36*, 2620–2627. [CrossRef]
5. Kumar, U.; Jain, V.K. Time series models (Grey-Markov, Grey Model with rolling mechanism and singular spectrum analysis) to forecast energy consumption in India. *Energy* **2010**, *35*, 1709–1716. [CrossRef]
6. Bianco, V.; Scarpa, F.; Tagliafico, L.A. Long term outlook of primary energy consumption of the Italian thermoelectric sector: Impact of fuel and carbon prices. *Energy* **2015**, *87*, 153–164. [CrossRef]
7. Topalli, A.K.; Erkmen, I. A hybrid learning for neural networks applied to short term load forecasting. *Neurocomputing* **2003**, *51*, 495–500. [CrossRef]
8. Kandil, N.; Wamkeue, R.; Saad, M.; Georges, S. An efficient approach for short term load forecasting using artificial neural networks. *Int. J. Electr. Power Energy Syst.* **2006**, *28*, 525–530. [CrossRef]
9. Beccali, M.; Cellura, M.; Brano, V.L.; Marvuglia, A. Forecasting daily urban electric load profiles using artificial neural networks. *Energy Convers. Manag.* **2004**, *45*, 2879–2900. [CrossRef]
10. Topalli, A.K.; Cellura, M.; Erkmen, I.; Topalli, I. Intelligent short-term load forecasting in Turkey. *Electr. Power Energy Syst.* **2006**, *28*, 437–447. [CrossRef]
11. Pai, P.F.; Hong, W.C. Forecasting regional electricity load based on recurrent support vector machines with genetic algorithms. *Electr. Power Syst. Res.* **2005**, *74*, 417–425. [CrossRef]
12. Hong, W.C. Electric load forecasting by seasonal recurrent SVR (support vector regression) with chaotic artificial bee colony algorithm. *Energy* **2011**, *36*, 5568–5578. [CrossRef]
13. Yesilbudak, M.; Sagiroglu, S.; Colak, I. A new approach to very short term wind speed prediction using k-nearest neighbor classification. *Energy Convers Manag.* **2013**, *69*, 77–86. [CrossRef]
14. Peng, H.; Liu, F.; Yang, X. A hybrid strategy of short term wind power prediction. *Renew. Energy* **2013**, *50*, 590–595. [CrossRef]
15. An, X.; Jiang, D.; Liu, C.; Zhao, M. Wind farm power prediction based on wavelet decomposition and chaotic time series. *Expert Syst Appl.* **2011**, *38*, 11280–11285. [CrossRef]
16. Lei, Y.; Lin, J.; He, Z.; Zuo, M.J. A review on empirical mode decomposition in fault diagnosis of rotating machinery. *Mech. Syst. Signal. Process.* **2013**, *35*, 108–126. [CrossRef]
17. Wong, P.; Xu, Q.; Vong, C.; Wong, H. Rate-dependent hysteresis modeling and control of a piezostage using online support vector machine and relevance vector machine. *IEEE Trans. Ind. Electron.* **2012**, *59*, 1988–2001. [CrossRef]
18. Wang, Z.; Liu, L. Sensitivity prediction of sensor based on relevance vector machine. *J. Inf. Comput. Sci.* **2012**, *9*, 2589–2597.
19. Hong, W.C. *Intelligent Energy Demand Forecasting*; Springer: London, UK, 2013.
20. Huang, B.; Kunoth, A. An optimization based empirical mode decomposition scheme. *J. Comput. Appl. Math.* **2013**, *240*, 174–183. [CrossRef]
21. Fan, G.; Qing, S.; Wang, H.; Shi, Z.; Hong, W.C.; Dai, L. Study on apparent kinetic prediction model of the smelting reduction based on the time series. *Math. Probl. Eng.* **2012**, *2012*. [CrossRef]
22. Bhusana, P.; Chris, T. Improving prediction of exchange rates using Differential EMD. *Expert Syst. Appl.* **2013**, *40*, 377–384.
23. An, X.; Jiang, D.; Zhao, M.; Liu, C. Short-term prediction of wind power using EMD and chaotic theory. *Commun. Nonlinear Sci. Numer. Simul.* **2012**, *17*, 1036–1042. [CrossRef]
24. Huang, N.E.; Shen, Z. A new View of nonliner water waves: The Hilbert spectrum. *Rev. Fluid Mech.* **1999**, *31*, 417–457. [CrossRef]
25. Fan, G.; Peng, L.-L.; Hong, W.-C.; Sun, F. Electric load forecasting by the SVR model with differential empirical mode decomposition and auto regression. *Neurocomputing* **2016**, *173*, 958–970. [CrossRef]
26. Wu, P.F.; Gao, L.Q.; Zou, D.X.; Li, S. An improved particle swarm optimization algorithm for reliability problems. *ISA. Trans.* **2011**, *50*, 71–81. [CrossRef] [PubMed]
27. Alfi, A. Particle swarm optimization algorithm with dynamic inertia weight for online parameter identification applied to Lorenz chaotic system. *Int. J. Innov. Comput. Inf.* **2012**, *8*, 1191–1203.
28. Alfi, A.; Modares, H. System identification and control using adaptive particle swarm optimization. *Appl. Math. Model.* **2011**, *35*, 1210–1221. [CrossRef]

29. Alfi, A. PSO with adaptive mutation and inertia weight and its application in parameter estimation of dynamic systems. *Acta Autom. Sin.* **2011**, *37*, 541–549. [CrossRef]

30. Wu, X.J.; Huang, Q.; Zhu, X.J. Thermal modeling of a solid oxide fuel cell and micro gas turbine hybrid power system based on modified LS-SVM. *Int. J. Hydrogen Energy.* **2011**, *36*, 885–892. [CrossRef]

31. Liu, X.Y.; Shao, C.; Ma, H.F.; Liu, R.X. Optimal earth pressure balance control for shield tunneling based on LS-SVM and PSO. *Autom. Constr.* **2011**, *29*, 321–327. [CrossRef]

32. Sun, J.; Feng, B.; Xu, W.B. Particle swarm optimization with particles having quantum behavior. In Proceedings of the IEEE Proceedings of Congress on Evolutionary Computation, New Orleans, LA, USA, 19–23 June 2004; pp. 325–331.

33. Sun, J.; Xu, W.B.; Feng, B. A global search strategy of quantum-behaved particle swarm optimization. In Proceedings of the Cybernetics and Intelligent Systems Proceedings of the 2004 IEEE Conference, Singapore, 4–9 May 2004; pp. 111–116.

34. Vapnik, V. *The Nature of Statistical Learning Theory*; Springer-Verlag: New York, NY, USA, 1995.

35. Bergh, F.V.D. An Analysis of Particle Swarm Optimizers. Ph.D. Thesis, University of Pretoria, Pretoria, South Africa, 2001.

36. Sun, J.; Fang, W.; Wu, X.J.; Palade, V.; Xu, W.B. Quantum-behaved particle swarm optimization: Analysis of individual particle behavior and parameter selection. *Evol. Comput.* **2012**, *20*, 349–393. [CrossRef] [PubMed]

37. Sun, J.; Fang, W.; Palade, V.; Wu, X.J.; Xu, W.B. Quantum-behaved particle swarm optimization with gaussian distributed local attractor point. *Appl. Math. Comput.* **2011**, *218*, 3763–3775. [CrossRef]

38. Sun, T.H. Applying particle swarm optimization algorithm to roundness measurement. *Expert Syst. Appl.* **2009**, *36*, 3428–3438. [CrossRef]

39. Xi, M.; Sun, J.; Xu, W. An improved quantum-behaved particle swarm optimization algorithm with weighted mean best position. *Appl. Math. Comput.* **2008**, *205*, 751–759. [CrossRef]

40. Coelho, L.S. An efficient particle swarm approach for mixed-integer programming in reliability—Redundancy optimization applications. *Reliab. Eng. Syst. Saf.* **2009**, *94*, 830–837. [CrossRef]

41. Fang, Y.; Zhao, Y.D.; Ke, M.; Zhao, X.; Iu, H.H.; Wong, K.P. Quantum inspired particle swarm optimization for power system operations considering wind power uncertainty and carbon tax in Australia. *IEEE Trans. Ind. Inform.* **2012**, *8*, 880–888.

42. Ko, C.N.; Jau, Y.M.; Jeng, J.T. Parameter estimation of chaotic dynamical systems using quantum-behaved particle swarm optimization based on hybrid evolution. *J. Inf. Sci. Eng.* **2015**, *31*, 675–689.

43. Ko, C.N.; Chang, Y.P.; Wu, C.J. An orthogonal-array-based particle swarm optimizer with nonlinear time-varying evolution. *Appl. Math. Comput.* **2007**, *191*, 272–279. [CrossRef]

44. Ratnaweera, A.; Halgamuge, S.K.; Watson, H.C. Self-organizing hierarchical particle swarm optimizer with time-varying acceleration coefficients. *IEEE Transactions on Evol. Comput.* **2004**, *8*, 240–255. [CrossRef]

45. Gao, J. Asymptotic properties of some estimators for partly linear stationary autoregressive models. *Commun. Statist. Theory Meth.* **1995**, *24*, 2011–2026. [CrossRef]

46. Taylor, J.W. Short-term load forecasting with exponentially weighted methods. *IEEE Trans. Power Syst.* **2012**, *27*, 458–464. [CrossRef]

47. Che, J.; Wang, J.; Wang, G. An adaptive fuzzy combination model based on self-organizing map and support vector regression for electric load forecasting. *Energy* **2012**, *37*, 657–664. [CrossRef]

48. Lowry, R. *Concepts & Applications of Inferential Statistics*; Vassar College: New York, NY, USA, 2011.

energies

MDPI

Article

Forecasting Crude Oil Price Using EEMD and RVM with Adaptive PSO-Based Kernels

Taiyong Li [1,2,*], Min Zhou [1,3], Chaoqi Guo [1], Min Luo [1], Jiang Wu [1], Fan Pan [4], Quanyi Tao [5] and Ting He [6,*]

1   School of Economic Information Engineering, Southwestern University of Finance and Economics, 55 Guanghuacun Street, Chengdu 610074, China; minzhousc@gmail.com (M.Z.); 214081203004@2014.swufe.edu.cn (C.G.); 41311040@2013.swufe.edu.cn (M.L.); wuj_t@swufe.edu.cn (J.W.)
2   Institute of Chinese Payment System, Southwestern University of Finance and Economics, 55 Guanghuacun Street, Chengdu 610074, China
3   School of Computer Science, Civil Aviation Flight University of China, Guanghan 618307, China
4   College of Electronics and Information Engineering, Sichuan University, 24 South Section 1, Yihuan Road, Chengdu 610065, China; popfan229@gmail.com
5   Huaan Video Technology Co., Ltd., Building 6, 399 Western Fucheng Avenue, Chengdu 610041, China; quanyitaoai@gmail.com
6   Department of Viral Vaccine, Chengdu Institute of Biological Products Co., Ltd., China National Biotech Group, 379 Section 3, Jinhua Road, Chengdu 610023, China
*   Correspondence: litaiyong@gmail.com (T.L.); cdhet@163.com (T.H.); Tel.: +86-28-8709-2220 (T.L.); +86-28-8441-8854 (T.H.)

Academic Editor: Wei-Chiang Hong
Received: 30 October 2016; Accepted: 25 November 2016; Published: 1 December 2016

Abstract: Crude oil, as one of the most important energy sources in the world, plays a crucial role in global economic events. An accurate prediction for crude oil price is an interesting and challenging task for enterprises, governments, investors, and researchers. To cope with this issue, in this paper, we proposed a method integrating ensemble empirical mode decomposition (EEMD), adaptive particle swarm optimization (APSO), and relevance vector machine (RVM)—namely, EEMD-APSO-RVM—to predict crude oil price based on the "decomposition and ensemble" framework. Specifically, the raw time series of crude oil price were firstly decomposed into several intrinsic mode functions (IMFs) and one residue by EEMD. Then, RVM with combined kernels was applied to predict target value for the residue and each IMF individually. To improve the prediction performance of each component, an extended particle swarm optimization (PSO) was utilized to simultaneously optimize the weights and parameters of single kernels for the combined kernel of RVM. Finally, simple addition was used to aggregate all the predicted results of components into an ensemble result as the final result. Extensive experiments were conducted on the crude oil spot price of the West Texas Intermediate (WTI) to illustrate and evaluate the proposed method. The experimental results are superior to those by several state-of-the-art benchmark methods in terms of root mean squared error (RMSE), mean absolute percent error (MAPE), and directional statistic (Dstat), showing that the proposed EEMD-APSO-RVM is promising for forecasting crude oil price.

Keywords: ensemble empirical mode decomposition (EEMD); particle swarm optimization (PSO); relevance vector machine (RVM); kernel methods; crude oil price; energy forecasting

1. Introduction

It was reported by British Petroleum (BP) that fossil fuels accounted for 86% of primary energy demand in 2014 and remain the dominant source of energy powering the global economy, with almost 80% of total energy supply in 2035. Among fossil fuels, crude oil is and will be the most important energy source, accounting for almost 29% of total energy supply in 2035 [1], and plays a vital role in all economies. In light of the importance of crude oil for the global economy, many enterprises, governments, investors, and researchers have devoted great efforts to building models to predict its price and volatility. However, due to its complexity, the price of oil can be easily affected by many factors, such as supply and demand, speculation activities, competition from providers, technique development, geopolitical conflicts, and wars [2–4]. All of these factors make the crude oil price nonlinear, nonstationary, and fluctuate with high volatility. For example, the West Texas Intermediate (WTI) crude oil price reached the peak of 145.31 USD per barrel in July 2008. However, the price drastically dropped to 30.28 USD per barrel, with about an 80% decrease from the peak at the end of 2008 because of the financial crisis. With economic recovery, the price rose above 113 USD per barrel in April 2011, and then sharply declined below 27 USD per barrel in February 2016 for changes of supply and demand, and for some political reasons.

A wide variety of models have emerged to predict crude oil price over the past decades, which could be roughly classified into two categories: (1) statistical and econometric models; (2) artificial intelligence (AI) models. Typically, statistical and econometric models include random walk model (RWM), error correction models (ECM), grey model (GM), vector autoregressive (VAR) models, autoregressive integrated moving average (ARIMA), and generalized autoregressive conditional heteroskedasticity (GARCH) family models. For instance, Hooper et al. [5] and Murat et al. [6] studied the performance of RWM in predicting crude oil price. The study of Baumeister and Kilin showed that a VAR model outperformed some compared methods in terms of accuracy when applied to forecasting crude oil price [7]. Xiang and Zhang analyzed and predicted monthly Brent crude oil price by ARIMA, and showed that model ARIMA(1,1,1) achieved good results [8]. As one of the most popular time series methods, ARIMA has been widely used as a benchmark in forecasting crude oil price by many scholars [4,9–11]. GARCH is another widely used method for forecasting crude oil price. Morana exploited the GARCH properties of the Brent crude oil price volatility and developed a semiparametric model based on the bootstrap approach to predict crude oil price [12]. Arouri applied an extended GARCH model to forecast the conditional volatility of crude oil price with structural breaks [13]. Mohammadi and Su applied ARIMA and GARCH to forecast the conditional mean and volatility of weekly crude oil price in several markets [14]. Since these statistical and econometric models are built on the assumption that crude oil price is linear and stationary, it is hard for them to predict nonlinear and nonstationary crude oil price with high performance.

As far as AI methods, artificial neural network (ANN) and support vector machine (SVM) have been widely used for predicting crude oil price. Shambora and Rossiter used an ANN model with moving average crossover inputs to forecast the future price of crude oil, and the results showed the superiority of ANN when compared with RWMs [15]. Mirmirani and Li compared VAR and ANN with genetic algorithm (GA) in forecasting crude oil price; the experimental results indicated that ANN with GA noticeably outperformed VAR [16]. Azadeh et al. compared ANN with fuzzy regression (FR) in forecasting long-term oil price in noisy, uncertain, and complex environments, and they concluded that ANN considerably outperformed FR in terms of mean absolute percentage error (MAPE) [17]. Tang and Zhang put forward a multiple wavelet recurrent neural network (MW-RNN) model for forecasting crude oil price, where wavelet and ANN were applied to capture multiscale data characteristics and to predict crude oil price at different scales, respectively. The proposed model could achieve high prediction accuracy [18]. Haidar et al. utilized a three-layer feedforward neural network to forecast short-term crude oil price [19]. SVM, first proposed by Vapnik [20], is a very popular supervised learning algorithm that can be applied to both classification and regression. The SVM for regression is also know as support vector regression (SVR). Xie et al. proposed an SVM-based method for crude oil price forecasting, and the results indicated that SVM outperformed ARIMA and back propagation neural network (BPNN) [21]. Li and Ge presented a novel model integrating ϵ-SVR and dynamic correction factor for forecasting crude oil price [11]. Some scholars studied the optimization of kernel types and/or kernel parameters in SVM for oil price forecasting [22,23]. Least squares support vector machine (LSSVM) [24]—an extension of SVM with less training time—has also been used in crude oil price forecasting [25]. Generally speaking, since the above-mentioned AI models can capture the nonlinear and nonstationary characteristics of crude oil price, these models are superior to the statistical and econometric models.

Owing to its highly complex characteristics of nonlinearity and nonstationarity, achieving satisfactory predictive accuracy on the raw crude oil price series is still a challenging task, although many attempts have been made. In recent years, a novel "decomposition and ensemble" framework has demonstrated its superiority in forecasting time series, which decomposes a complex times series into a few simple components, predicts each component individually, and finally ensembles all predicted values as final result [4,9,26–29]. The simple components can effectively preserve some features of complex raw data from different perspectives, and each of them can be independently handled with relatively simple methods. The challenging task of forecasting crude oil price from the complex raw data is divided into several relatively easy subtasks of forecasting each component. Therefore, this framework is effective for forecasting crude oil price. For example, Yu et al. proposed a model based on empirical mode decomposition (EMD) and ANN to predict WTI and Brent crude oil price, and the results demonstrated the attractiveness of the proposed model [9]. Yu et al. also proposed a novel model based on ensemble EMD (EEMD) and extended extreme learning machine (EELM) to predict the crude oil price of WTI [4,30]. Zhang et al. put forward a novel hybrid model with EEMD, LSSVM, particle swarm optimization (PSO), and GARCH to predict crude oil price, where LSSVM with parameters optimized by PSO and GARCH were used to forecast nonlinear and time-varying components by EEMD, respectively [26]. Tang et al. integrated complementary EEMD (CEEMD) and EELM to forecast crude oil price [27]. In addition, Fan et al. used independent component analysis (ICA) to decompose the crude oil price time series into three independent components, and then constructed three SVR models to predict the components respectively, and finally used SVR again to integrate the results by the former three SVRs as final price [31].

Relevance vector machine (RVM) [32]—a kernel-trick machine learning method that uses Bayesian inference—has attracted much attention from researchers in both classification and regression in recent years [33–39]. The main advantages of RVM over SVM are the absence of a regularizing parameter, and the ability to use non-Mercer kernels, probabilistic output, and sparsity formulation. The kernel types and kernel parameters are still crucial in RVM. For example, Fei et al. and Wang et al. studied the performance of wavelet kernel in RVM [40,41]. The authors used composite kernels to identify nonlinear systems [42]. Psorakis et al. investigated the sparsity and accuracy of multi-class multi-kernel RVMs [43]. To improve the performance of RVM, some evolutionary algorithms were applied to optimize the weight of single kernel or kernel parameters. Fei and He used an extended PSO to optimize the weight and parameters in a combined kernel by a radial basic function (RBF) kernel and a polynomial kernel for state prediction of bearing [44], and Zhang et al. used a similar method to predict the capacity of Lithium-Ion Batteries [45]. GA, artificial bee colony algorithm (ABC), and ant colony optimization algorithms (ACO) were also applied to optimize kernel parameters in RVM [46–48]. Regarding time series analysis, RVM has been successful in detecting seizure in electroencephalogram (EEG) signals [49] and forecasting stock index [50], exchange rate [51], nonlinear hydrological time series [52], wind speed [47,53], and the price of electricity [54]. These applications show the superiority of RVM in time series forecasting. According to the existing literature, there was little research on crude oil price forecasting by RVM.

As a popular decomposition method, EEMD has advantages over other methods : (1) it can be used to decompose nonlinear and nonstationary signals into several IMFs and one residue; (2) the IMFs by EEMD are obtained adaptively and represent local features of the signal; (3) unlike Fourier and wavelet transforms, EEMD does not need a basis function for decomposition; and (4) it needs only two parameters (the number of ensemble and the standard deviation of Gaussian white noise). Therefore, it can be seen that the incorporation of EEMD as the decomposition method, RVM as the prediction method, and addition as the ensemble method might achieve good accuracy of crude oil price forecasting, following the "decomposition and ensemble" framework. Based on the framework, the original difficult task of forecasting crude oil price is divided into several relatively easy subtasks of forecasting each component individually. Since EEMD decomposes the raw crude oil price into a set of components and the raw price equals the sum of all the components, simple addition might be a good choice to ensemble all predicted results from components as the final result. Although EEMD and kernel methods have succeeded in forecasting time series, most of the existing studies used a fixed type of kernel to predict every component by EEMD, ignoring the characteristics of the data. In fact, each component has its own characteristics. For example, the residue reflects the trend of original signal, while the first intrinsic mode function (IMF) reflects the highest frequency [30]. It is more appropriate to adaptively select kernel types and kernel parameters for each component by its own characteristics [55]. To cope with this issue, this research aims to propose a novel method integrating EEMD, adaptive PSO (APSO), and RVM—namely, EEMD-APSO-RVM—to predict crude oil price following the "decomposition and ensemble" framework. Specifically, the raw price was decomposed into several components. Then, for each component, RVM with a combined kernel where weights and parameters of single kernels were optimized by an extended PSO was applied to predict its target value. Finally, the predicted values of all components were aggregated as final predicted crude oil price. Compared to the basic "decomposition and ensemble" framework, the proposed EEMD-APSO-RVM improves the accuracy of crude oil price forecasting in three aspects: (1) it uses EEMD instead of some other decomposition methods to decompose the raw time series into several components that can better represent the characteristics of the data; (2) it applies RVM to forecast each component because of its good predictive capabilities; and (3) it proposes APSO to adaptively optimize the weights and parameters of the single kernels in the combined kernel of RVM. The main contributions of this work are three-fold: (1) we proposed an EEMD-APSO-RVM to predict crude oil price. To the best of our knowledge, it is the first time that RVM has been applied to forecasting crude oil price; (2) an extended PSO was employed to simultaneously optimize kernel types and kernel parameters for RVM, resulting in an optimal kernel for the specified component by EEMD; (3) extensive experiments were conducted on WTI crude oil price, and the results demonstrated that the proposed EEMD-APSO-RVM method is promising for forecasting crude oil price. Accordingly, the novelty of this paper can be described as : (1) it introduces RVM to forecasting crude oil price for the first time; and (2) an adaptive PSO is proposed to optimize the weights and parameters of kernels in RVM to improve the accuracy of crude oil price forecasting.

The remainder of this paper is organized as follows. Section 2 describes the formulation process of the proposed EEMD-APSO-RVM method in detail. Experimental results are reported and analyzed in Section 3. Finally, Section 4 concludes this paper.

2. Methodology

The decomposition and ensemble framework has three steps; i.e., decomposition, individual prediction, and ensemble prediction. In this section, the overall formulation process of EEMD-APSO-RVM is presented. Firstly, the related EEMD, PSO, and RVM are briefly introduced individually in Sections 2.1–2.3. Secondly, the adaptive PSO for parameters optimization in RVM is described in Section 2.4. Finally, the EEMD-APSO-RVM algorithm is formulated, and the corresponding steps are described in detail in Section 2.5.

2.1. Ensemble Empirical Mode Decomposition

Ensemble empirical mode decomposition (EEMD) is an extended version of empirical mode decomposition (EMD) developed to overcome the drawback of the so-called "mode mixing" problem in the latter [30,56]. Contrary to traditional decomposition methodologies, EEMD is an empirical, direct, intuitive, and self-adaptive methodology that can decompose nonlinear and nonstationary time series into components (several IMFs and one residue), with each component having a length equal to the original signal. Since it was proposed, it has been widely applied to complex system analysis, showing its superiority in forecasting time series.

The main idea of EEMD is to perform EMD many times on the time series, given a number of Gaussian white noises to obtain a set of IMFs, and then the ensemble average of corresponding IMFs is treated as the final decomposed results. The main steps of EEMD are as follows:

Step 1: Specify the number of ensemble M and the standard deviation of Gaussian white noises σ, with $i = 0$;

Step 2: $i = i + 1$; Add a Gaussian white noise $n_i(t) \sim N(0, \sigma^2)$ to crude oil price series $X(t)$ to construct a new series $X_i(t)$, as follows:

$$X_i(t) = X(t) + n_i(t). \tag{1}$$

Step 3: Decompose $X_i(t)$ into m IMFs $c_{ij}(t)(j = 1, \ldots, J)$ and a residue $r_i(t)$, as follows:

$$X_i(t) = \sum_{j=1}^{m} c_{ij}(t) + r_i(t), \tag{2}$$

where c_{ij} is the j-th IMF in the i-th trial, and J is the number of IMFs, determined by the size of crude oil price series N with $J = \lfloor log_2 N \rfloor - 1$ [30].

Step 4: If $i < M$, go to Step 2 to perform EMD again; otherwise, go to Step 5;

Step 5: Calculate the average of corresponding IMFs of M trials as final IMFs:

$$c_j(t) = \frac{1}{M} \sum_{i=1}^{M} c_{ij}(t), i = 1, \ldots, M; j = 1, \ldots, J. \tag{3}$$

Once the EEMD completes, the original time series can be expressed as the sum of J IMFs and a residue, as follows:

$$X(t) = \sum_{j=1}^{J} c_j(t) + r_{J,t}, \tag{4}$$

where $r_{J,t}$ is the final residue. Now, the issue of forecasting original time series becomes the new issue of forecasting each component decomposed by EEMD.

2.2. Particle Swarm Optimization

Particle swarm optimization (PSO)—firstly proposed by Eberhart and Kennedy—is an evolutionary computation algorithm that uses the velocity-displacement model through iteration to simulate swarm intelligence [57]. The algorithm initializes with a group of random particles in space of D dimensions, and each particle—representing a potential solution—is assigned a randomized velocity to change its position, searching for the optimal solution. In each iteration, the particles keep track of the local best solution p_l and the global best solution p_g to decide the flight speed and distance accordingly.

The ith particle has a position vector and a velocity vector in D dimensional space, described as $p_i = (p_{i1}, p_{i2}, \ldots, p_{iD})$ and $v_i = (v_{i1}, v_{i2}, \ldots, v_{iD})$, and the optimum locations achieved by the ith particle and population are also described as $p_{li} = (p_{li1}, p_{li2}, \ldots, p_{liD})$ and $p_g = (p_{g1}, p_{g2}, \ldots, p_{gD})$, respectively. The formulas to update the speed and position of the dth dimension of the ith particle are as follows, respectively:

$$v_{id}(t+1) = w v_{id}(t) + c_1 r_1 (p_{lid} - p_{id}(t)) + c_2 r_2 (p_{gd} - p_{id}(t)) \tag{5}$$

$$p_{id}(t+1) = p_{id}(t) + v_{id}(t+1)) \tag{6}$$

where t is the current number of iteration, w is inertia weight, c_1 and c_2 are nonnegative accelerate constants, and r_1 and r_2 are random in the range of [0,1].

PSO is good at real optimization. Therefore, in this research, we use PSO to optimize the weight and parameters in each single kernel for the combined kernel in RVM.

2.3. Relevance Vector Machine

Relevance vector machine (RVM)—put forward by Tipping [32]—can be applied to both regression and classification. Since forecasting crude oil price is related to regression, here we give a brief review of RVM for regression only. Readers can refer to [32] for more details on RVM.

Given a set of samples $\{x_i, t_i\}_{i=1}^N$, where $x_i \in R^d$ are d-dimensional vectors as inputs and $t_i \in R$ are real values as targets, and assuming that $t_i = y(x_i; w) + \epsilon_i$ with $\epsilon_i \sim N(0, \sigma^2)$, the RVM model for regression can be formulated as:

$$t = y(x; w) = \sum_{i=1}^N w_i K(x, x_i) + w_0, \tag{7}$$

where $K(x, x_i)$ is a kernel function on x and x_i, and w_i is the weight of the kernel. Then, for a sample i, the conditional probability of the target is as follows:

$$p(t_i|x_i) = N(t_i|y(x_i; w), \sigma^2). \tag{8}$$

Assuming that the samples $\{x_i, t_i\}_{i=1}^N$ are independently generated, the likelihood of all the samples can be defined as follows:

$$\begin{aligned} p(t|w, \sigma^2) &= \prod_{i=1}^N N(t_i|y(x_i; w), \sigma^2) \\ &= (2\pi\sigma^2)^{-\frac{N}{2}} exp(-\frac{||t - \Phi w||^2}{2\sigma^2}), \end{aligned} \tag{9}$$

where Φ is a design matrix having the size $N \times (N+1)$ with $\Phi = [\phi(x_1), \phi(x_2), \ldots, \phi(x_N)]^T$, wherein each component is the vector of the response of kernel function associated with the sample x_n as $\phi(x_n) = [1, K(x_n, x_1), K(x_n, x_2), \ldots, K(x_n, x_N)]^T$. It may cause over-fitting if we implement maximum-likelihood estimation for w and σ^2 directly, because the size of training samples is almost the same as the size of parameters. To overcome this, Tipping imposed a constraint on weights w from a Bayesian perspective, as follows [32]:

$$p(w|\alpha) = \prod_{i=0}^N N(w_i|0, \alpha_i^{-1}), \tag{10}$$

where α is an $N+1$ vector named hyperparameters. With the prior on weights, for all unknown samples, the posterior can be computed from the proceeds of Bayes inference as:

$$p(w, \alpha, \sigma^2|t) = \frac{p(t|w, \alpha, \sigma^2) \times p(w, \alpha, \sigma^2)}{p(t)}. \tag{11}$$

For a given input point x_*, the predictive distribution of the corresponding target t_* can be written as:

$$p(t_*|t) = \int p(t_*|w, \alpha, \sigma^2) p(w, \alpha, \sigma^2|t) dw d\alpha d\sigma^2. \tag{12}$$

It is difficult to directly compute the posterior $p(w, \alpha, \sigma^2 | t)$ in Equation (11). Instead, Tipping further decomposes it as follows:

$$p(w, \alpha, \sigma^2 | t) = p(w | t, \alpha, \sigma^2) p(\alpha, \sigma^2 | t). \tag{13}$$

The computation of $p(w, \alpha, \sigma^2 | t)$ is now becoming the computation of two items: $p(w | t, \alpha, \sigma^2)$ and $p(\alpha, \sigma^2 | t)$. The posterior distribution over weights can be written from Bayes's rule:

$$
\begin{aligned}
p(w | t, \alpha, \sigma^2) &= \frac{p(t | w, \sigma^2) p(w | \alpha)}{p(t | \alpha, \sigma^2)} \\
&= (2\pi\sigma^2)^{-\frac{N+1}{2}} |\sum|^{-\frac{1}{2}} exp(-\frac{(w - \mu)^T \sum^{-1} (w - \mu)}{2}),
\end{aligned} \tag{14}
$$

where the posterior covariance and mean are as follows, respectively,

$$\sum = (\beta \Phi^T \Phi + A)^{-1}, \tag{15}$$

$$\mu = \beta \sum \Phi^T t, \tag{16}$$

with $\beta = \sigma^{-2}$ and $A = diag(\alpha_0, \alpha_1, \dots, \alpha_N)$, respectively.

As far as the second item at right hand side of Equation (13), it can be decomposed as:

$$p(\alpha, \sigma^2 | t) \propto p(t | \alpha, \sigma^2) p(\alpha) p(\sigma^2) \propto p(t | \alpha, \sigma^2). \tag{17}$$

Therefore, the learning process of RVM is now transformed to maximizing Equation (18) with respect to the hyperparameters α and σ^2:

$$
\begin{aligned}
p(t | \alpha, \sigma^2) &= \int p(t | w, \sigma^2) p(w | \alpha) dw \\
&= (2\pi)^{-\frac{N}{2}} |\sigma^2 I + \Phi A^{-1} \Phi^T|^{-\frac{1}{2}} exp(-\frac{t^T (\sigma^2 I + \Phi A^{-1} \Phi^T)^{-1} t}{2}),
\end{aligned} \tag{18}
$$

where I is an identity matrix.

By simply setting the derivatives of Equation (18) to zero, we can obtain the re-estimation equations on α and σ^2 as follows, respectively:

$$\alpha_i^{new} = \frac{1 - \alpha_i \sum_{ii}}{\mu_i^2}, \tag{19}$$

$$(\sigma^2)^{new} = \frac{||t - \Phi\mu||^2}{N - \sum_i (1 - \alpha_i \sum_{ii})}. \tag{20}$$

With the iteration, the optimal values of α and σ^2—termed as α_{MP} and σ_{MP}^2 respectively—can be achieved by maximizing Equation (18).

Finally, for the given input point t_*, the predictive result can be computed as follows:

$$p(t_* | t, \alpha_{MP}, \sigma_{MP}^2) = \int p(t_* | w, \sigma_{MP}^2) p(w | t, \alpha_{MP}, \sigma_{MP}^2) dw = N(t_* | y_*, \sigma_*^2), \tag{21}$$

where $y_* = \mu^T \phi(x_*)$ and $\sigma_*^2 = \sigma_{MP}^2 + \phi(x_*)^T \sum \phi(x_*)$.

The kernel function in RVM plays a crucial role which significantly influences the performance of RVM. Therefore, it is important to select appropriate kernels according to the characteristics of the data instead of using a single fixed kernel. Some widely used single kernels include the linear kernel $K_{lin}(x_i, y_i) = x_i^T y_i$, the polynomial kernel $K_{poly}(x_i, y_i) = (a(x_i^T y_i) + b)^c$, the RBF kernel $K_{rbf}(x_i, y_i) = exp(-\frac{||x_i - y_i||^2}{2d})$ (here we use d to represent σ^2 for short), and the sigmoid kernel

$K_{sig}(x_i, y_i) = tanh(e(x_i^T y_i) + f)$. Among the kernels, the parameters $a - f$ usually need to be specified by users. In this paper, we integrate the above-mentioned four kernels into a combined kernel for RVM, which can be represented as:

$$K_{comb}(x_i, y_i) = \lambda_1 K_{lin}(x_i, y_i) + \lambda_2 K_{poly}(x_i, y_i) + \lambda_3 K_{rbf}(x_i, y_i) + \lambda_4 K_{sig}(x_i, y_i), \tag{22}$$

where $\lambda_1 - \lambda_4$ are the weights for the four kernels that satisfy $\sum_{i=1}^{4} \lambda_i = 1$. In this way, each single kernel of the four kernels is a special case of the combined kernel. For example, when $\lambda_1 = \lambda_2 = \lambda_4 = 0$ and $\lambda_3 = 1$, the combined kernel degenerates to the RBF kernel. In the combined kernel, ten parameters $(\lambda_1, \lambda_2, \lambda_3, \lambda_4, a, b, c, d, e, $ and $f)$ need to be optimized.

2.4. Adaptive PSO for Parameter Optimization in RVM

For a specific problem, it is hard to set appropriate values for the parameters in the combined kernel in Equation (22) according to priori knowledge. PSO is a widely used real optimization algorithm that could be used in this case. However, in traditional PSO, the inertia weight for each particle in one generation is fixed, and it varies with the iteration—ignoring the difference among particles. Some varieties of PSO adaptively adjust the inertia wight of each particle based on one or more feedback parameters [58]. Ideally, the particles far from the global best particle should have larger inertia weight with more exploration ability, while the ones close to the global best particle should have smaller inertia weight with more exploitation ability. To cope with this issue, in this paper, an adaptive PSO (APSO) is proposed to optimize the parameters in RVM, which adaptively adjusts the inertia weight of each particle in an iteration according to the distance between the current particle and the global best particle.

Definition 1. *Distance between two particles. The distance between two particles p_i and p_j can be defined as:*

$$dist(p_i, p_j) = \sqrt{\sum_{k=1}^{d} (p_{ik} - p_{jk})^2 + (f(p_i) - f(p_j))^2}, \tag{23}$$

where d is the dimension of particle, and f is the fitness function. It is worth noting that each dimension in Equation (23) needs to be mapped into the same scale (e.g., [0,1]) in order for the computation to make sense. According to this definition, the distance between two particles has three properties: (1) $dist(p_i, p_j) = dist(p_j, p_i)$; (2) $dist(p_i, p_i) = 0$; (3) $dist(p_i, p_k) + dist(p_k, p_j) \geq dist(p_i, p_j)$.

Definition 2. *Average distance of population. The average distance of the population can be defined as:*

$$mdist = \frac{2 \sum_{i=1}^{N} \sum_{j=1}^{i-1} dist(p_i, p_j)}{N(N-1)}, \tag{24}$$

where N is the total number of particles in a swarm.

In this paper, we propose an adaptive strategy to adjust the inertia weight for one particle p_i in the t-th iteration by Equation (25):

$$w_{t,i} = \begin{cases} w_{min} + \frac{w_{max} - w_{min}}{T} t, & dist(p_i, p_g) > mdist \\ w_{min} + \frac{w_{max} - w_{min}}{mdist} dist(p_i, p_g), & dist(p_i, p_g) \leq mdist \end{cases} \tag{25}$$

where T is the number of total iterations, p_g is the global best particle, and w_{max} and w_{min} are the maximal and minimal inertia weights specified by users, respectively. The main idea of Equation (25) is to adjust the inertia weight of each particle adaptively according to its distance from the global best particle. If the current particle is far from the global best particle, it uses traditional inertia weight. Otherwise, it adaptively adjusts its inertia weight according to its distance to the global best particle.

The model using APSO to optimize the parameters of the combined kernel in RVM—called APSO-RVM—can be presented as:

Step 1: Setting parameters. Set the following parameters for running APSO, population size P, maximal iteration times T, the maximal and minimal inertia weights w_{max} and w_{min}, the range of the ten parameters to be optimized;

Step 2: Encoding. Encode the ten parameters into a particle (vector) $p_i = (p_{i1}, p_{i2}, \ldots, p_{i10})$ to represent $\lambda_1, \lambda_2, \lambda_3, \lambda_4, a, b, c, d, e$, and f accordingly;

Step 3: Defining the fitness function. The fitness function is defined by root mean square error (RMSE):

$$f(p_i) = \sqrt{\frac{1}{N} \sum_{n=1}^{N} (y_i - \phi(x_i, p_i))^2} \, , \tag{26}$$

where N is the size of training samples, y_i is the true target of the input x_i, and $\phi(x_i, p_i)$ is the predicted target associated with x_i and the parameter p_i;

Step 4: Initializing. Set $t = 0$; randomly generate initial speed and position for each particle; use the value of particle p_i to compose the kernel for RVM in Equation (22), and then evaluate each particle; p_i is selected as p_{li}, while the particle with the optimal fitness is selected as p_g;

Step 5: Updating speed and position. Set $t = t + 1$; calculate the inertia weight using Equation (25), and update the speed and position according to Equations (5) and (6), respectively;

Step 6: Evaluating particles. Evaluate each particle by fitness function;

Step 7: Updating the historical best particle, if necessary. If $f(p_i) \leq f(p_{li})$, then $p_{li} = p_i$;

Step 8: Updating the global best particle, if necessary. If $f(p_i) \leq f(p_g)$, then $p_g = p_i$;

Step 9: Judging whether the iteration terminates or not. If $t \leq T$, go to Step 5. Otherwise, stop the iteration and output p_{gb} as the optimized parameters for the combined kernel in RVM. The optimal RVM predictor is obtained at this point.

The APSO is based on the framework of PSO, and the main improvement lies in that each particle has its own inertia weight according to its distance from the global best particle. In this paper, the APSO is applied to adaptively searching the optimal weights and parameters of the single kernels for the combined kernel in RVM to predict crude oil price.

2.5. The Proposed EEMD–APSO–RVM Model

Following the framework of "decomposition and ensemble", a three-stage methodology that integrates ensemble empirical mode decomposition (EEMD), adaptive particle swarm optimization (APSO), and relevance vector machine (RVM)—termed EEMD-APSO-RVM—can be formulated for forecasting crude oil price. As shown in Figure 1, the proposed EEMD-APSO-RVM generally consists of three main stages:

Stage 1: Decomposition. The original crude oil price series $x_t, (t = 1, 2, \ldots, T)$ is decomposed into with $J = \lfloor log_2 T \rfloor - 1$ intrinsic mode function (IMF) components $c_{j,t}, (j = 1, 2, \ldots, J)$ and one residue component $r_{N,t}$ using EEMD;

Stage 2: Individual forecasting. RVM with the combined kernel optimized by APSO is used to forecast each component in Stage 1 independently, resulting in the predicted values of IMFs $\hat{c}_{j,t}$ and that of the residue $\hat{r}_{N,t}$, respectively;

Stage 3: Ensemble forecasting. The final predicted results $\hat{x}_t$ can be obtained by simply adding the predicted results of all IMF components and the residue; i.e., $\hat{x}_t = \sum_{j=1}^{J} \hat{c}_{j,t} + \hat{r}_{N,t}$.

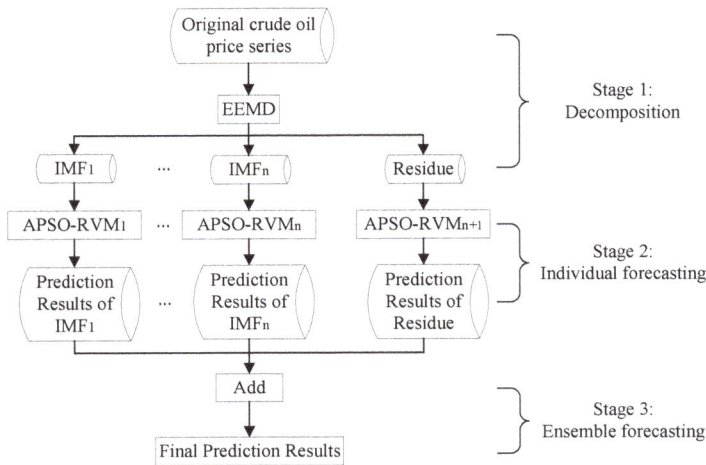

Figure 1. Flowchart for the proposed method. APSO: adaptive particle swarm optimization; EEMD: ensemble empirical mode decomposition; IMF: intrinsic mode function; RVM: relevance vector machine.

The proposed EEMD-APSO-RVM is one of the typical strategies of "divide and conquer". The complicated question of forecasting the original crude oil price is transformed to several questions of forecasting relatively simple components independently. The EEMD-APSO-RVM adopts a combined kernel that integrates four commonly used kernels. Furthermore, the weights and parameters in the combined kernel are adaptively optimized by an extension of PSO. The EEMD-APSO-RVM decomposes the crude oil price into several IMFs and one residue for forecasting individually, instead of using the nonlinear and nonstationary raw data as the input to a single forecasting method; this can improve the forecasting accuracy, because the individual forecasting is a relatively easy task. The kernel-trick RVM has the ability to accurately predict time series such as wind speed and electricity price, which will benefit crude oil price forecasting. The APSO adaptively optimizes the parameters in the kernel, trying to find the optimal kernel to improve the forecasting results. All these attributes make it possible for the EEMD-APSO-RVM to improve the accuracy of crude oil price forecasting.

3. Numerical Example

To demonstrate the performance of the proposed EEMD-APSO-RVM, in this paper, we select the crude oil price of West Texas Intermediate (WTI) as experimental data, as described in Section 3.1. The evaluation criteria are introduced in Section 3.2. Section 3.3 gives the parameter settings and data preprocessing for the experiments, and in Section 3.4, the experimental results are reported. We further analyse the robustness and running time of the proposed method in Section 3.5. Finally, some interesting findings can be obtained from the experimental study.

3.1. Data Description

The crude oil price of WTI can be accessed from the US energy information administration (EIA) [59]. We use the daily close price covering the period of 2 January 1986 to 12 September 2016, with 7743 observations in total for experiments. Among the observations, the first 6194 from 2 January 1986 to 21 July 2010 are treated as training samples, while the remaining 1549 from 22 July 2010 to 12 September 2016 are for testing—accounting for 80% and 20% of total observations, respectively.

We conduct h-step-ahead predictions with horizon $h = 1, 3, 6$ in this study. Given a time series $x_t, (t = 1, 2, \ldots, T)$, the h-step-ahead prediction for x_{t+h} can be formulated as:

$$\hat{x}_{t+h} = f(x_{t-(l-1)}, x_{t-(l-2)}, \ldots, x_{t-1}, x_t), \tag{27}$$

where $\hat{x}_{t+h}$ is the h-step-ahead predicted value at time t, x_t is the true value at time t, and l is the lag orders.

3.2. Evaluation Criteria

The root mean squared error (RMSE), the mean absolute percent error (MAPE), and the directional statistic (D_{stat}) are selected to evaluate the performance of the proposed method. With the true value x_t and the predicted value $\hat{x}_t$ at time t, RMSE is defined as:

$$RMSE = \sqrt{\frac{1}{N} \sum_{t=1}^{N} (x_t - \hat{x}_t)^2}, \tag{28}$$

where N is the number of testing observations. Note that the RMSE here has the same meaning as Equation (26), where the predicted value is represented by $\phi(x_i, p_i)$.

As another evaluation criteria for prediction accuracy, MAPE is defined as:

$$MAPE = \frac{1}{N} \sum_{t=1}^{N} |\frac{x_t - \hat{x}_t}{x_t}|. \tag{29}$$

In addition, Dstat measures the ability to forecast the direction of price movement, which is defined as:

$$D_{stat} = \frac{1}{N} \sum_{t=1}^{N} \alpha_t \times 100\%, \tag{30}$$

where $\alpha_t = 0$ if $(\hat{x}_{t+1} - x_t)(x_{t+1} - x_t) < 0$; otherwise, $\alpha_t = 1$.

An ideal forecasting method should achieve low RMSE, low MAPE, and high D_{stat}.

3.3. Experimental Settings

In order to evaluate the performance of the proposed methods, some state-of-the-art models were selected as benchmarks to compare with the EEMD-APSO-RVM. In the decomposition stage, we select EMD as a benchmark. In the prediction stage, the compared models include one classical statistical method (ARIMA) and two popular AI models (LSSVR and ANN). In addition, RVM with a single kernel (RVMlin, RVMpoly, RVMrbf, RVMsig) and RVM with a combined kernel from the former four single kernels optimized by standard PSO (PSO-RVM) are also independently employed in this stage. Therefore, we have eight single methods (PSO-RVM, RVMlin, RVMpoly, RVMrbf, RVMsig, ANN, LSSVR, and ARIMA) to compare with APSO-RVM, and fifteen ensemble methods (EEMD-PSO-RVM, EEMD-RVMlin, EEMD-RVMpoly, EEMD-RVMrbf, EEMD-RVMsig, EEMD-ANN, EEMD-LSSVR, EMD-PSO-RVM, EMD-APSO-RVM, EMD-RVMlin, EMD-RVMpoly, EMD-RVMrbf, EMD-RVMsig, EMD-ANN, and EMD-LSSVR) to compare with EEMD-APSO-RVM. All methods are shown in Table 1.

The parameters for APSO are listed in Table 2. The standard PSO uses the same parameters as APSO. Note that to guarantee $\sum_{i=1}^{4} \lambda_i = 1$, we simply map the values in particles to new values to be applied to the combined kernel with $\lambda_j' = \frac{\lambda_j}{\sum_{i=1}^{4} \lambda_i}$. For b, we use $b = round(b)$ to get an integer as the exponent. Following some previous work [4,27], we apply RBF kernel in LSSVR and use grid search to find the optimal γ and σ^2 in the range of $\{2^k, k = -4, -3, \ldots, 12\}$ and $\{2^k, k = -4, -3, \ldots, 12\}$, respectively. For ANN, we use a back propagation neural network and set ten as the number of hidden nodes. The iteration times of ANN was set to 10,000. For the parameters in single RVM-related predictors (i.e., $a - f$), we search the best parameters in the same ranges as those in APSO (listed in Table 2) with an interval of 0.2, excepting that c varies with an interval of 1 and d varies in $\{2^k, k = -4, -3, \ldots, 12\}$. We use the Akaike information criterion (AIC) [60] to determine the ARIMA parameters (p-d-q). We also set the lag orders in Equation (27) to six, as analysed in [61].

Table 1. Descriptions of all the methods in the experiments. ANN: artificial neural network; PSO: particle swarm optimization.

Type	Name	Descriptions		
		Decomposition	Forecasting	Ensemble
Single	APSO-RVM	-	RVM with a combined kernel optimized by APSO	-
	PSO-RVM	-	RVM with a combined kernel optimized by standard PSO	-
	RVMlin	-	RVM with a linear kernel	-
	RVMpoly	-	RVM with a polynomial kernel	-
	RVMrbf	-	RVM with a radial basic function kernel	-
	RVMsig	-	RVM with a sigmoid kernel	-
	ANN	-	Back propagation neural network	-
	LSSVR	-	Least squares support vector regression	-
	ARIMA	-	Autoregressive integrated moving average	-
Ensemble	EEMD-APSO-RVM	EEMD	RVM with a combined kernel optimized by APSO	Addition
	EEMD-PSO-RVM	EEMD	RVM with a combined kernel optimized by standard PSO	Addition
	EEMD-RVMlin	EEMD	RVM with a linear kernel	Addition
	EEMD-RVMpoly	EEMD	RVM with a polynomial kernel	Addition
	EEMD-RVMrbf	EEMD	RVM with a radial basic function kernel	Addition
	EEMD-RVMsig	EEMD	RVM with a sigmoid kernel	Addition
	EEMD-ANN	EEMD	Back propagation neural network	Addition
	EEMD-LSSVR	EEMD	Least squares support vector regression	Addition
	EMD-APSO-RVM	EMD	RVM with a combined kernel optimized by APSO	Addition
	EMD-PSO-RVM	EMD	RVM with a combined kernel optimized by standard PSO	Addition
	EMD-RVMlin	EMD	RVM with a linear kernel	Addition
	EMD-RVMpoly	EMD	RVM with a polynomial kernel	Addition
	EMD-RVMrbf	EMD	RVM with a radial basic function kernel	Addition
	EMD-RVMsig	EMD	RVM with a sigmoid kernel	Addition
	EMD-ANN	EMD	Back propagation neural network	Addition
	EMD-LSSVR	EMD	Least squares support vector regression	Addition

Table 2. Parameters for APSO.

Description	Symbol	Range / Value
Population size	P	20
Maximal iterations	T	40
Particle dimension	D	10
Maximal, minimal inertia weight	w_{max}, w_{min}	0.9, 0.4
Accelerate constants	c_1, c_2	1.49, 1.49
Kernel weight	$\lambda_1, \lambda_2, \lambda_3, \lambda_4$	[0, 1]
Coefficient in K_{poly}	a	[0, 2]
Constant in K_{poly}	b	[0, 10]
Exponent in K_{poly}	c	[1, 4]
Width in K_{rbf}	d	$[2^{-4}, 2^{12}]$
Coefficient in K_{sig}	e	[0, 4]
Constant in K_{sig}	f	[0, 8]

Regarding ensemble models, we firstly add white noise with a standard deviation of 0.15 to the original crude oil price, and then set 100 as the number of ensembles in EEMD. The decomposition results of the original crude oil price by EEMD is shown in Figure 2, with 11 IMFs and one residue.

To set up the stage for a fair comparison, we applied the Min–Max Normalization (as shown in Equation (31)) for all of the data:

$$x_{norm} = (x - x_{min})/(x_{max} - x_{min}), \tag{31}$$

where x_{min} and x_{max} are the minimal and maximal values for one dimension in data, respectively, and x_{norm} and x are the normalized and the original values, respectively. It is clear that the normalization maps the original values to the range $[0, 1]$. Conversely, after obtaining the predicted value from the normalized data $\hat{x}_{norm}$, the corresponding expected predicted value $\hat{x}$ in original scale can be computed as:

$$\hat{x} = x_{min} + (x_{max} - x_{min}) * \hat{x}_{norm}. \tag{32}$$

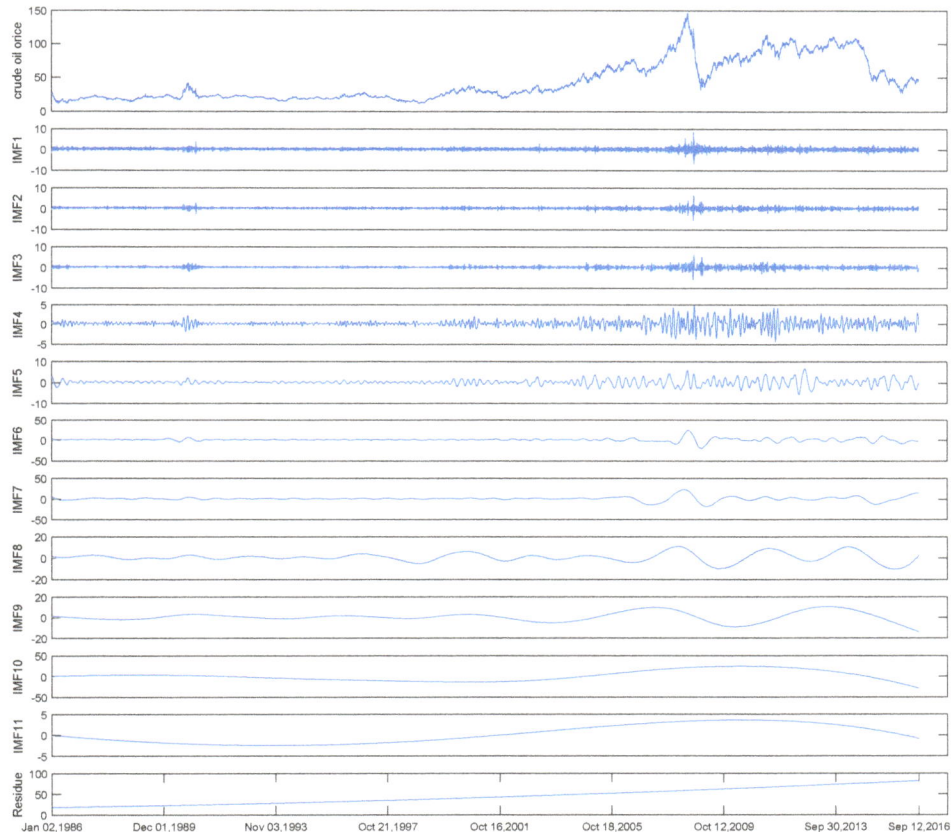

Figure 2. The IMF and residue components by EEMD.

All of the experiments were conducted by Matlab 8.6 (Mathworks, Natick, MA, USA) on a 64-bit Windows 7 (Microsoft, Redmond, WA, USA) with 32 GB memory and 3.4 GHz I7 CPU.

3.4. Results and Analysis

3.4.1. Results of Single Models

We firstly evaluate the single models (i.e., APSO-RVM, PSO-RVM, RVMlin, RVMpoly, RVMrbf, RVMsig, ANN, LSSVR, and ARIMA) in terms of MAPE, RMSE, and Dstat, as shown in Figures 3–5. From these results, it can be concluded that the proposed APSO-RVM might be the most powerful single model among all the single models in forecasting crude oil price.

33

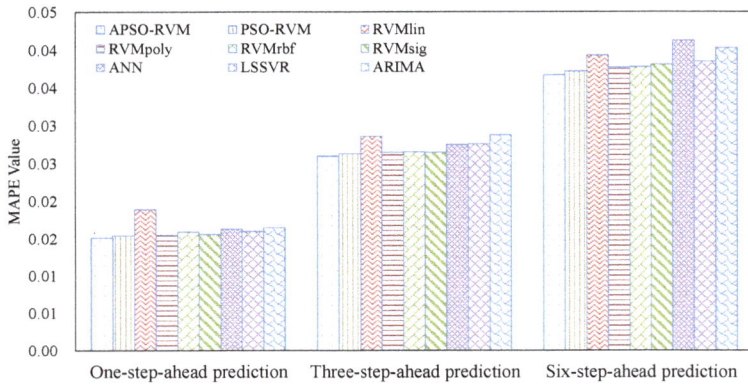

Figure 3. Mean absolute percentage error (MAPE) by different single methods.

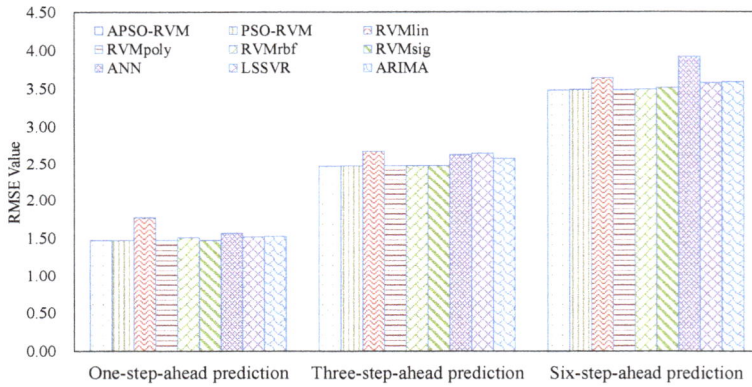

Figure 4. Root mean square error (RMSE) by different single methods.

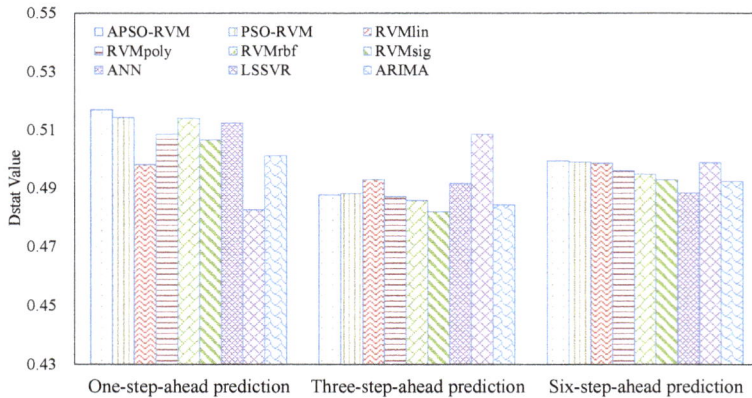

Figure 5. Dstat by different single methods.

The MAPE value by the APSO-RVM is the lowest amongst the nine single models at all horizons, followed by PSO-RVM, RVMpoly, RVMrbf, and RVMsig. The performances of the latter four models are quite alike, except that the MAPE value by RVMrbf at horizon one is slightly high. RVMlin achieves the highest values at horizon one and horizon three, and the third highest value at horizon six, showing its poor performance in forecasting crude oil price. The possible reason for this is that the crude oil price data is not linearly separable. The results by the state-of-the-art AI benchmark models (ANN and LSSVR) are very close at horizon one and horizon three. However, LSSVR outperforms ANN at horizon six. The statistical model—ARIMA—ranks sixth in all cases. This is probably because, as a typical linear model, it is difficult for ARIMA to accurately forecast crude oil price due, to its nonlinearity and nonstationarity.

As far as RMSE, the prediction accuracy of APSO-RVM is still ranked first among all of the compared benchmark models in all cases, although it is very close to the corresponding result by PSO-RVM. For the RVM model with a single kernel, RVMpoly, RVMrbf, and RVMsig achieve very close results, which are slightly higher than that of APSO-RVM, followed by RVMlin with the poorest results at horizon one and horizon three, and the second poorest result at horizon six among all methods. ANN, LSSVR, and ARIMA achieve very similar RMSE values at each horizon, except ANN underperforms LSSVR and ARIMA at horizon six.

From the perspective of directional accuracy, all of the models produce quite similar results, ranging from 0.48 to 0.52. It can be easily seen that none of the models can be proven to be better than the others. In spite of its leading performance in terms of MAPE and RMSE, APSO-RVM does not significantly outperform other models at all horizons regarding Dstat. The APSO-RVM ranks first at horizon one, fifth at horizon three, and first with slight advantages at horizon six. It is interesting that LSSVR ranks first at horizon three and second at horizon six, but it ranks last at horizon one. Another interesting finding is that the values of seven out of nine models at horizon six are higher than those at horizon three. Therefore, the performance of single models is not stable when forecasting the direction of crude oil price.

From the results by single models, it can be seen that none of the methods can consistently outperform others in all cases in terms of MAPE, RMSE, and Dstat. Another interesting finding is that many methods achieve very close results in most cases, although the APSO-RVM is better than others in eight out of nine cases. In addition, all of the results by the methods are undesirable, even for the best result. For example, the results of Dstat by all methods were between 0.48 and 0.52, which tends to guessing randomly, making it unpractical. All of these findings show that it is a difficult task to accurately forecast crude oil price using the nonlinear and nonstationary raw price. The main reason might be that the single models have their limitations in achieving high accuracy because of the complexity of crude oil price. Hence, in this work, we develop a novel "decomposition and ensemble" method to improve the performance of single models in forecasting crude oil price.

3.4.2. Results of Ensemble Models

Regarding the ensemble models (i.e., EEMD-APSO-RVM, EEMD-PSO-RVM, EEMD-RVMlin, EEMD-RVMpoly, EEMD-RVMrbf, EEMD-RVMsig, EEMD-ANN, EEMD-LSSVR, EMD-APSO-RVM, EMD-PSO-RVM, EMD-RVMlin, EMD-RVMpoly, EMD-RVMrbf, EMD-RVMsig, EMD-ANN, and EMD-LSSVR), Figures 6–8 show the corresponding results in terms of MAPE, RMSE, and Dstat. From these results, it can be easily seen that the proposed EEMD-APSO-RVM is the best model that achieves the lowest MAPE value, the lowest RMSE value, and the highest Dstat value at each horizon.

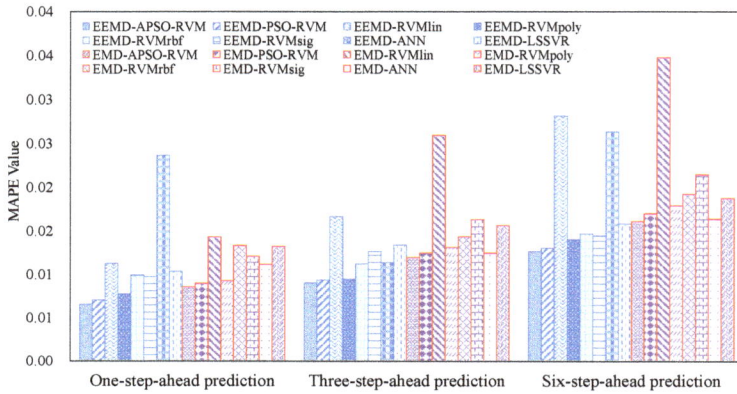

Figure 6. MAPE by different ensemble methods.

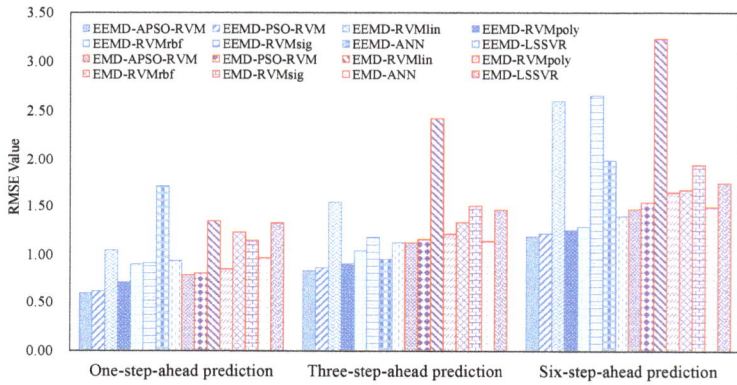

Figure 7. RMSE by different ensemble methods.

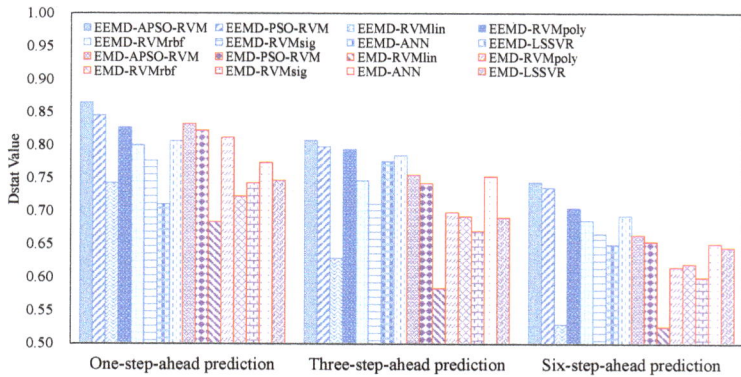

Figure 8. Dstat by different ensemble methods.

At each horizon, the MAPE value of EEMD-APSO-RVM ranks first among all models, being far lower than that of many other ensemble models. At the same time, the MAPE value of EMD-APSO-RVM also ranks first among all of the EMD-related methods. It shows the superiority of APSO-RVM in forecasting crude oil price. Accordingly, the EEMD-PSO-RVM and the EMD-PSO-RVM rank the second among EEMD-related and EMD-related methods, respectively, at each horizon, with slightly worse results than those of counterpart EEMD-APSO-RVM and EMD-APSO-RVM. EEMD-RVMpoly ranks third in terms of MAPE at each horizon, and EEMD-RVMsig and EEMD-RVMrbf are slightly worse than EEMD-RVMpoly, but are still better than many other methods. Among the EEMD-RVM family methods, EEMD-RVMlin is the poorest model, and it always ranks last at three horizons when compared with other EEMD-RVM-related models. It is clear that RVM with a combined kernel outperforms RVMs with a single kernel. Regarding ANN and LSSVR, it is interesting that ANN underperforms LSSVR twice with EEMD, while the first always outperforms the latter with EMD. For these two AI models, it is difficult to judge which is superior to the other, since they are both parameter-sensitive and it is difficult for traditional methods to find their optimal parameters. From the perspective of decomposition algorithms, it can be found that the ensemble methods with EEMD as decomposition method are much better than their counterpart methods with EMD, except for ANN at horizon one and horizon six, showing that the EEMD is a more effective decomposition method in time series analysis. Furthermore, EEMD-APSO-RVM significantly decreases the MAPE values when compared with the single APSO-RVM method, demonstrating the effectiveness of the decomposition method for forecasting performance.

Focusing on the RMSE values (shown in Figure 7), findings similar to those of MAPE can be obtained. EEMD-APSO-RVM still ranks first amongst all benchmark models, with 0.59, 0.83, and 1.18 at horizon one, horizon three, and horizon six, respectively. The results of EEMD-APSO-RVM are far less than those by any other models, except EEMD-PSO-RVM has slightly worse results at corresponding horizons. This further confirms that the proposed EEMD-APSO-RVM is effective for forecasting crude oil price. Most ensemble methods obviously outperform their corresponding single method. This is mainly attributed to the fact that EMD or EEMD can remarkably improve the prediction power of the models. Generally speaking, the ensemble methods with EEMD have better results than their corresponding methods with EMD, due to the good performance of EEMD on data analysis.

As far as Dstat (shown in Figure 8), all the values by ensemble models are higher than 0.525, and are quite different from the results of single models (as shown in Figure 5), where the highest value is less than 0.520. This demonstrates that the "decomposition and ensemble" framework can notably improve the performance of directional prediction. At each horizon, the proposed EEMD-APSO-RVM achieves the highest Dstat value (0.86, 0.81, and 0.74 at horizon one, horizon three, and horizon six, respectively), showing its superiority over all other methods. Similarly, EMD-APSO-RVM also outperforms all other EMD-related models at each horizon. The poorest results were usually achieved by RVM models with linear kernel, except that the EEMD-RVMlin obtains the second poorest value at horizon one, further demonstrating that the components from crude oil price are not linearly separable.

3.5. Analysis of Robustness and Running Time

Although EEMD-APSO-RVM succeeds in forecasting crude oil price, it has disadvantages. First, since the PSO uses many random values in the evolutionary process, it is hard for it to reproduce the experiments with the exact solutions. Second, it is time-consuming for the EEMD-APSO-RVM to find the optimal parameters and to compute the combined kernel.

To evaluate the robustness and stableness of the proposed EEMD-APSO-RVM, we repeated the experiments 10 times and report the results in terms of means and standard deviations (std.) of MAPE, RMSE, and Dstat in Table 3. It can be seen that the standard deviations of MAPE and Dstat are far less than 0.01, and at the same time, the standard deviation in each case is lower than 5% of corresponding mean. For RMSE, the standard deviations are slightly higher than those of MAPE and Dstat. However, even the poorest standard deviation in terms of RMSE is still less than 6% of the corresponding mean. The results show that EEMD-APSO-RVM is quite stable and robust for forecasting crude oil price.

Table 3. Statistical results of running the experiment ten times by the EEMD-APSO-RVM (mean ± std.).

Horizon	MAPE	RMSE	Dstat
One	0.0065 ± 0.0001	0.5905 ± 0.0110	0.8643 ± 0.0032
Three	0.0091 ± 0.0001	0.8324 ± 0.0340	0.8062 ± 0.0037
Six	0.0126 ± 0.0003	1.1843 ± 0.0702	0.7028 ± 0.0028

In the training phase of the EEMD-APSO-RVM, to find the optimal parameters for the combined kernel, many particles need to be evaluated by fitness function, which is time-consuming. It takes about 10 h to train a model at one horizon in our experimental environment (Matlab 8.6 on a 64-bit Windows 7 with 32 GB memory and 3.4 GHz I7 CPU), while it takes only about 3 s to test the 1549 samples with the optimized parameters. In practice, the testing time plays a more important role than the training time, because the training phase is usually completed with off-line data and it runs only once. Therefore, the time consumed by the EEMD–APSO–RVM is acceptable.

3.6. Summarizations

From the above discussions, some interesting findings can be obtained, as follows:

(1) Due to nonlinearity and nonstationarity, it is difficult for single models to accurately forecast crude oil price.
(2) The RVM has a good ability to forecast crude oil price. Even with a single kernel, SVM may outperform LSSVM, ANN, and ARIMA in many cases.
(3) The combined kernel can further improve the accuracy of RVM. PSO can be applied to optimize the weights and parameters of the single kernels for the combined kernel in RVM. In this case, the proposed APSO outperforms the traditional PSO.
(4) The EEMD-related methods achieve better results than the counterpart EMD-related methods, showing that EEMD is more suitable for decomposing crude oil price.
(5) With the benefits of EEMD, APSO, and RVM, the proposed ensemble EEMD–APSO–RVM significantly outperforms any other compared models listed in this paper in terms of MAPE, RMSE, and Dstat. At the same time, it is a stable and effective forecasting method in terms of robustness and running time. These all show that the EEMD–APSO–RVM is promising for crude oil price forecasting.

4. Conclusions

This paper proposes a novel model integrating EEMD, adaptive PSO, and RVM (namely EEMD-APSO-RVM) for forecasting crude oil price based on the "decomposition and ensemble" framework. In the decomposition phase, we used EEMD to decompose the raw crude oil price into components of several IMFs and one residue. In the single forecasting phase, we utilized RVM with a combined kernel optimized by an adaptive PSO to forecast each component individually. Finally, the predicted results of all components were aggregated by simple addition. To validate the EEMD-APSO-RVM, eight other single benchmark models and fifteen ensemble models were employed to compare the forecasting results of the crude oil spot price of WTI at three different horizons in terms of MAPE, RMSE, and Dstat. To the best of our knowledge, it is the first time that RVM with combined kernels have been applied to forecasting crude oil price. It can be concluded from the extensive experimental results that: (1) the APSO-RVM outperforms other single models in most cases; (2) the components by decomposition can better represent the characteristics of crude oil price than raw data. Furthermore, EEMD is superior to EMD for decomposition; and (3) the EEMD-APSO-RVM achieves satisfactory results in all cases, showing that it is promising for forecasting crude oil price.

In the future, the work could be extended in two aspects: (1) studying multiple kernel RVM to improve the performance on forecasting crude oil price; and (2) applying the EEMD-APSO-RVM to forecasting other time series of energy, such as wind speed and electricity price.

Acknowledgments: This work was supported in part by the Major Research Plan of the National Natural Science Foundation of China (Grant No. 91218301), the Fundamental Research Funds for the Central Universities (Grant No. JBK130503) and the Natural Science Foundation of China (Grant No. 71473201). It was also supported by the Collaborative Innovation Center for the Innovation and Regulation of Internet-Based Finance, Southwestern University of Finance and Economics.

Author Contributions: Taiyong Li and Ting He are the principal investigators of this work. They proposed the forecasting method and designed the experiments. Jiang Wu, Fan Pan and Quanyi Tao provided professional guidance. Taiyong Li, Min Zhou, Chaoqi Guo and Min Luo performed the experiments and analyzed the data. Taiyong Li wrote the manuscript. All authors have revised and approved the final manuscript.

Conflicts of Interest: The authors declare no conflict of interest.

References

1. British Petroleum. 2016 Energy Outlook. 2016. Available online: https://www.bp.com/content/dam/bp/pdf/energy-economics/energy-outlook-2016/bp-energy-outlook-2016.pdf (accessed on 28 August 2016).
2. Wang, Y.; Wei, Y.; Wu, C. Detrended fluctuation analysis on spot and futures markets of West Texas Intermediate crude oil. *Phys. A* **2011**, *390*, 864–875.
3. He, K.J.; Yu, L.; Lai, K.K. Crude oil price analysis and forecasting using wavelet decomposed ensemble model. *Energy* **2012**, *46*, 564–574.
4. Yu, L.A.; Dai, W.; Tang, L. A novel decomposition ensemble model with extended extreme learning machine for crude oil price forecasting. *Eng. Appl. Artif. Intell.* **2016**, *47*, 110–121.
5. Hooper, V.J.; Ng, K.; Reeves, J.J. Quarterly beta forecasting: An evaluation. *Int. J. Forecast.* **2008**, *24*, 480–489.
6. Murat, A.; Tokat, E. Forecasting oil price movements with crack spread futures. *Energy Econ.* **2009**, *31*, 85–90.
7. Baumeister, C.; Kilian, L. Real-time forecasts of the real price of oil. *J. Bus. Econ. Statist.* **2012**, *30*, 326–336.
8. Xiang, Y.; Zhuang, X.H. Application of ARIMA model in short-term prediction of international crude oil price. *Adv. Mater. Res.* **2013**, *798*, 979–982.
9. Yu, L.A.; Wang, S.Y.; Lai, K.K. Forecasting crude oil price with an EMD-based neural network ensemble learning paradigm. *Energy Econ.* **2008**, *30*, 2623–2635.
10. He, A.W.; Kwok, J.T.; Wan, A.T. An empirical model of daily highs and lows of West Texas Intermediate crude oil prices. *Energy Econ.* **2010**, *32*, 1499–1506.
11. Li, S.; Ge, Y. Crude Oil Price Prediction Based on a Dynamic Correcting Support Vector Regression Machine. *Abstr. Appl. Anal.* **2013**, *2013*, 528678.
12. Morana, C. A semiparametric approach to short-term oil price forecasting. *Energy Econ.* **2001**, *23*, 325–338.
13. Arouri, M.E.H.; Lahiani, A.; Lévy, A.; Nguyen, D.K. Forecasting the conditional volatility of oil spot and futures prices with structural breaks and long memory models. *Energy Econ.* **2012**, *34*, 283–293.
14. Mohammadi, H.; Su, L. International evidence on crude oil price dynamics: Applications of ARIMA-GARCH models. *Energy Econ.* **2010**, *32*, 1001–1008.
15. Shambora, W.E.; Rossiter, R. Are there exploitable inefficiencies in the futures market for oil? *Energy Econ.* **2007**, *29*, 18–27.
16. Mirmirani, S.; Li, H.C. A comparison of VAR and neural networks with genetic algorithm in forecasting price of oil. *Adv. Econom.* **2004**, *19*, 203–223.
17. Azadeh, A.; Moghaddam, M.; Khakzad, M.; Ebrahimipour, V. A flexible neural network-fuzzy mathematical programming algorithm for improvement of oil price estimation and forecasting. *Comput. Ind. Eng.* **2012**, *62*, 421–430.
18. Tang, M.; Zhang, J. A multiple adaptive wavelet recurrent neural network model to analyze crude oil prices. *J. Econ. Bus.* **2012**, *64*, 275–286.
19. Haidar, I.; Kulkarni, S.; Pan, H. Forecasting model for crude oil prices based on artificial neural networks. In Proceedings of the IEEE International Conference on Intelligent Sensors, Sensor Networks and Information Processing (ISSNIP 2008), Sydney, Australia, 15–18 December 2008; pp. 103–108.
20. Vapnik, V. *The Nature of Statistical Learning Theory*; Springer: New York, NY, USA, 2013.

21. Xie, W.; Yu, L.; Xu, S.; Wang, S. A new method for crude oil price forecasting based on support vector machines. In *Computational Science—ICCS 2006*; Springer: Berlin/Heidelberg, Germany, 2006; pp. 444–451.

22. Chiroma, H.; Abdulkareem, S.; Abubakar, A.I.; Herawan, T. Kernel functions for the support vector machine: Comparing performances on crude oil price data. In *Recent Advances on Soft Computing and Data Mining*; Springer: Basel, Switzerland, 2014; pp. 273–281.

23. Guo, X.; Li, D.; Zhang, A. Improved support vector machine oil price forecast model based on genetic algorithm optimization parameters. *AASRI Procedia* **2012**, *1*, 525–530.

24. Suykens, J.A.; Vandewalle, J. Least squares support vector machine classifiers. *Neural Process. Lett.* **1999**, *9*, 293–300.

25. Yu, Y.L.; Li, W.; Sheng, D.R.; Chen, J.H. A novel sensor fault diagnosis method based on Modified Ensemble Empirical Mode Decomposition and Probabilistic Neural Network. *Measurement* **2015**, *68*, 328–336.

26. Zhang, X.Y.; Liang, Y.T.; Zhou, J.Z.; Zang, Y. A novel bearing fault diagnosis model integrated permutation entropy, ensemble empirical mode decomposition and optimized SVM. *Measurement* **2015**, *69*, 164–179.

27. Tang, L.; Dai, W.; Yu, L.; Wang, S. A novel CEEMD-based EELM ensemble learning paradigm for crude oil price forecasting. *Int. J. Inf. Technol. Decis. Mak.* **2015**, *14*, 141–169.

28. Yu, L.; Wang, Z.; Tang, L. A decomposition–ensemble model with data-characteristic-driven reconstruction for crude oil price forecasting. *Appl. Energy* **2015**, *156*, 251–267.

29. He, K.J.; Zha, R.; Wu, J.; Lai, K.K. Multivariate EMD-Based Modeling and Forecasting of Crude Oil Price. *Sustainability* **2016**, *8*, 387, doi:10.3390/su8040387.

30. Wu, Z.; Huang, N.E. Ensemble empirical mode decomposition: A noise-assisted data analysis method. *Adv. Adapt. Data Anal.* **2009**, *1*, 1–41.

31. Fan, L.; Pan, S.; Li, Z.; Li, H. An ICA-based support vector regression scheme for forecasting crude oil prices. *Technol. Forecast. Soc. Chang.* **2016**, *112*, 245–253.

32. Tipping, M.E. Sparse Bayesian learning and the relevance vector machine. *J. Mach. Learn. Res.* **2001**, *1*, 211–244.

33. Chen, S.; Gunn, S.R.; Harris, C.J. The relevance vector machine technique for channel equalization application. *IEEE Trans. Neural Netw.* **2002**, *12*, 1529–1532.

34. Kim, J.; Suga, Y.; Won, S. A new approach to fuzzy modeling of nonlinear dynamic systems with noise: Relevance vector learning mechanism. *IEEE Trans. Fuzzy Syst.* **2006**, *14*, 222–231.

35. Tolambiya, A.; Kalra, P.K. Relevance vector machine with adaptive wavelet kernels for efficient image coding. *Neurocomputing* **2010**, *73*, 1417–1424.

36. De Martino, F.; de Borst, A.W.; Valente, G.; Goebel, R.; Formisano, E. Predicting EEG single trial responses with simultaneous fMRI and Relevance Vector Machine regression. *Neuroimage* **2011**, *56*, 826–836.

37. Mehrotra, H.; Singh, R.; Vatsa, M.; Majhi, B. Incremental granular relevance vector machine: A case study in multimodal biometrics. *Pattern Recognit.* **2016**, *56*, 63–76.

38. Gupta, R.; Laghari, K.U.R.; Falk, T.H. Relevance vector classifier decision fusion and EEG graph-theoretic features for automatic affective state characterization. *Neurocomputing* **2016**, *174*, 875–884.

39. Kiaee, F.; Sheikhzadeh, H.; Mahabadi, S.E. Relevance Vector Machine for Survival Analysis. *IEEE Trans. Neural Netw. Learn. Syst.* **2016**, *27*, 648–660.

40. Fei, H.; Xu, J.W.; Min, L.; Yang, J.H. Product quality modelling and prediction based on wavelet relevance vector machines. *Chemom. Intell. Lab. Syst.* **2013**, *121*, 33–41.

41. Wang, F.; Gou, B.C.; Qin, Y.W. Modeling tunneling-induced ground surface settlement development using a wavelet smooth relevance vector machine. *Comput. Geotech.* **2013**, *54*, 125–132.

42. Camps-Valls, G.; Martinez-Ramon, M.; Rojo-Alvarez, J.L.; Munoz-Mari, J. Nonlinear system identification with composite relevance vector machines. *IEEE Signal. Process. Lett.* **2007**, *14*, 279–282.

43. Psorakis, I.; Damoulas, T.; Girolami, M.A. Multiclass Relevance Vector Machines: Sparsity and Accuracy. *IEEE Trans. Neural Netw.* **2010**, *21*, 1588–1598.

44. Fei, S.W.; He, Y. A Multiple-Kernel Relevance Vector Machine with Nonlinear Decreasing Inertia Weight PSO for State Prediction of Bearing. *Shock Vib.* **2015**, *2015*, doi:10.1155/2015/685979.

45. Zhang, Y.L.; Zhou, W.D.; Yuan, S.S. Multifractal Analysis and Relevance Vector Machine-Based Automatic Seizure Detection in Intracranial EEG. *Int. J. Neural Syst.* **2015**, *25*, 149–154.

46. Yuan, J.; Wang, K.; Yu, T.; Fang, M.L. Integrating relevance vector machines and genetic algorithms for optimization of seed-separating process. *Eng. Appl. Artif. Intell.* **2007**, *20*, 970–979.

47. Fei, S.W.; He, Y. Wind speed prediction using the hybrid model of wavelet decomposition and artificial bee colony algorithm-based relevance vector machine. *Int. J. Electr. Power Energy Syst.* **2015**, *73*, 625–631.

48. Liu, D.T.; Zhou, J.B.; Pan, D.W.; Peng, Y.; Peng, X.Y. Lithium-ion battery remaining useful life estimation with an optimized Relevance Vector Machine algorithm with incremental learning. *Measurement* **2015**, *63*, 143–151.

49. Zhang, Y.S.; Liu, B.Y.; Zhang, Z.L. Combining ensemble empirical mode decomposition with spectrum subtraction technique for heart rate monitoring using wrist-type photoplethysmography. *Biomed. Signal Process. Control* **2015**, *21*, 119–125.

50. Huang, S.C.; Wu, T.K. Combining wavelet-based feature extractions with relevance vector machines for stock index forecasting. *Expert Syst.* **2008**, *25*, 133–149.

51. Huang, S.C.; Hsieh, C.H. Wavelet-Based Relevance Vector Regression Model Coupled with Phase Space Reconstruction for Exchange Rate Forecasting. *Int. J. Innov. Comput. Inf. Control* **2012**, *8*, 1917–1930.

52. Liu, F.; Zhou, J.Z.; Qiu, F.P.; Yang, J.J.; Liu, L. Nonlinear hydrological time series forecasting based on the relevance vector regression. In Proceedings of the 13th International Conference on Neural Information Processing (ICONIP'06), Hong Kong, China, 3–6 October 2006; pp. 880–889.

53. Sun, G.Q.; Chen, Y.; Wei, Z.N.; Li, X.L.; Cheung, K.W. Day-Ahead Wind Speed Forecasting Using Relevance Vector Machine. *J. Appl. Math.* **2014**, *2014*, 437592, doi:10.1155/2014/437592.

54. Alamaniotis, M.; Bargiotas, D.; Bourbakis, N.G.; Tsoukalas, L.H. Genetic Optimal Regression of Relevance Vector Machines for Electricity Pricing Signal Forecasting in Smart Grids. *IEEE Trans. Smart Grid* **2015**, *6*, 2997–3005.

55. Zhang, X.; Lai, K.K.; Wang, S.Y. A new approach for crude oil price analysis based on empirical mode decomposition. *Energy Econ.* **2008**, *30*, 905–918.

56. Huang, N.E.; Shen, Z.; Long, S.R. A new view of nonlinear water waves: The Hilbert Spectrum 1. *Annu. Rev. Fluid Mech.* **1999**, *31*, 417–457.

57. Eberhart, R.C.; Kennedy, J. A new optimizer using particle swarm theory. In Proceedings of the Sixth International Symposium on Micro Machine and Human Science, New York, NY, USA, 4–6 October 1995; Volume 1, pp. 39–43.

58. Nickabadi, A.; Ebadzadeh, M.M.; Safabakhsh, R. A novel particle swarm optimization algorithm with adaptive inertia weight. *Appl. Soft Comput.* **2011**, *11*, 3658–3670.

59. EIA Website. Available online: http://www.eia.doe.gov (accessed on 20 September 2016).

60. Liu, H.; Tian, H.Q.; Li, Y.Q. Comparison of two new ARIMA-ANN and ARIMA-Kalman hybrid methods for wind speed prediction. *Appl. Energy* **2012**, *98*, 415–424.

61. Yu, L.; Zhao, Y.; Tang, L. A compressed sensing based AI learning paradigm for crude oil price forecasting. *Energy Econ.* **2014**, *46*, 236–245.

energies

MDPI

Article

Electric Load Forecasting Based on a Least Squares Support Vector Machine with Fuzzy Time Series and Global Harmony Search Algorithm

Yan Hong Chen [1], Wei-Chiang Hong [2,3,*], Wen Shen [1] and Ning Ning Huang [1]

[1] School of Information, Zhejiang University of Finance & Economics, Hangzhou 310018, China;
    zufe_graphics@163.com (Y.H.C.); hnn0702@163.com (W.S.); swing_w@163.com (N.N.H.)
[2] School of Economics & Management, Nanjing Tech University, Nanjing 211800, China
[3] Department of Information Management, Oriental Institute of Technology, 58 Sec. 2, Sichuan Road,
    Panchiao, Taipei 220, Taiwan
* Correspondence: samuelsonhong@gmail.com; Tel.: +886-2-7738-0145 (ext. 5316); Fax: +886-2-7738-6310

Academic Editor: Sukanta Basu
Received: 19 October 2015; Accepted: 21 January 2016; Published: 26 January 2016

Abstract: This paper proposes a new electric load forecasting model by hybridizing the fuzzy time series (FTS) and global harmony search algorithm (GHSA) with least squares support vector machines (LSSVM), namely GHSA-FTS-LSSVM model. Firstly, the fuzzy c-means clustering (FCS) algorithm is used to calculate the clustering center of each cluster. Secondly, the LSSVM is applied to model the resultant series, which is optimized by GHSA. Finally, a real-world example is adopted to test the performance of the proposed model. In this investigation, the proposed model is verified using experimental datasets from the Guangdong Province Industrial Development Database, and results are compared against autoregressive integrated moving average (ARIMA) model and other algorithms hybridized with LSSVM including genetic algorithm (GA), particle swarm optimization (PSO), harmony search, and so on. The forecasting results indicate that the proposed GHSA-FTS-LSSVM model effectively generates more accurate predictive results.

Keywords: electric load forecasting; least squares support vector machine (LSSVM); global harmony search algorithm (GHSA); fuzzy time series (FTS); fuzzy c-means (FCM)

1. Introduction

Load forecasting plays an important role in electric system planning and operation. In recent years, lots of researchers have studied the load forecasting problem and developed a variety of load forecasting methods. Load forecasting algorithms can be divided into three major categories: traditional methods, modern intelligent methods and hybrid algorithms [1]. The traditional method [1,2] mainly includes autoregressive (AR), autoregressive moving average (ARMA) [3], autoregressive integrated moving average (ARIMA) [4], semi-parametric [5], gray model [6,7], similar-day models [8], and Kalman filtering method [9]. Due to the theoretical limitations of the algorithms themselves, it is difficult to improve the forecasting accuracy using these forecasting approaches. For example, the ARIMA model is unable to capture the rapid changing process underlying the electric load from historical data pattern. The Kalman filter model cannot avoid the observation noise and the forecasting accuracy of the grey model will be reduced along with the increasing degree of discretiin of the data.

The intelligent methods mainly include artificial neural network (ANN) [10], fuzzy systems [11], knowledge based expert system (KBES) approach [12], wavelet analysis [13], support vector machine (SVM) [14], and so on. Knowledge-based expert system combines the knowledge and experience of numerous experts to maximize the experts' ability, but the method does not have self-learning ability. Besides, KBES is limited to the total amount of knowledge stored in the database and it is difficult to process any sudden change of the conditions [15]. The ANN has the ability of nonlinear approximation, self-learning, parallel processing and higher adaptive ability, however, it also has some problems, such as the difficulty of choosing parameters, and high computational complexity. SVM is a new machine learning method proposed by Cortes and Vapnik [16]. It is based on the principle of structural risk minimization (SRM) in statistical learning theory. The practical problems such as small sample, nonlinear, high dimension and local minimum point could be solved by the SVM via solving a convex quadratic programming (QP) problem. However, traditional SVM also has some shortcomings. For example, SVM cannot determine the input variables effectively and reasonably and it has slow convergence speed and poor forecasting results while suffering from strong random fluctuation time series. Compared with SVM, the least squares support vector machine (LSSVM), proposed by Suykens and Vandewalle [17], is an improved model of the original SVM. It has the following advantages, using equality constraints instead of the inequality in standard SVM, solving a set of linear equations instead of QP [13]. LSSVM has been widely applied to solve forecasting problems in many fields, such as stock index forecasting [18], credit rating forecasting [19], GPRS traffic forecasting [20], tax forecasting [21] and prevailing wind direction forecasting [22], and so on.

Fuzzy time series (FTS), as a significant quantitative forecasting model, has been broadly applied in electric load forecasting. There are lots of literatures focused on FTS related issues that are also involved in this paper [23–27]. Lee and Hong [23] proposed a new FTS approaches for the electric power load forecasting. Efendi *et al.* [24] discussed the fuzzy logical relationships used to determine the electric load forecast in the FTS modeling. Sadaei *et al.* [26] presented an enhanced hybrid method based on a sophisticated exponentially weighted fuzzy algorithm to forecast short-term load. FTS is often combined with other models for forecasting. For example, a new method for forecasting the TAIEX is presented based on FTS and SVMs [28].

In addition, various optimization algorithms are widely employed in LSSVM to improve its searching performance, such as genetic algorithm (GA) [29], particle swarm optimization (PSO) [30], harmony search algorithm (HSA) and artificial bee colony algorithm (ABC) [31]. All the optimization methods improve the efficiency of the model in some way. Although the single forecasting method can improve the forecasting accuracy in some aspects, it is more difficult to yield the desired accuracy in all electric load forecasting cases. Thus, via hybridizing two or more approaches, the hybrid model can combine the merits of two or more models, as proposed by researchers. A new hybrid forecasting method, namely ESPLSSVM, based on empirical mode decomposition, seasonal adjustment, PSO and LSSVM model is proposed in [32]. Hybridization of support vector regression (SVR) with chaotic sequence and EA is able to avoid solutions trapping into a local optimum and improve forecasting accuracy successfully [33]. Ghofrani *et al.* [34] proposed a hybrid forecasting framework by applying a new data preprocessing algorithm with time series and regression analysis to enhance the forecasting accuracy of a Bayesian neural network (BNN). A hybrid algorithm based on fuzzy algorithm and imperialist competitive algorithm (RHWFTS-ICA) is also developed [35], in which the fuzzy algorithm is refined high-order weighted. In this paper, the global harmony search algorithm (GHSA) is hybridized with LSSVM to optimize the parameters of LSSVM.

The rest of this paper consists three sections: the proposed method GHSA-FTS-LSSVM, including FTS model, fuzzy c-means clustering (FCS) algorithm, GA, global harmony search and least squares SVM, is introduced in Section 2; a numerical example is illustrated in Section 3; and conclusions are discussed in Section 4.

2. Methodology of Global Harmony Search Algorithm-Fuzzy Time Series-Least Squares Support Vector Machines Model

2.1. Least Squares Support Vector Machine Model

LSSVM is a kind of supervised learning model which is widely used in both classification problems and regression analysis. Comparing with SVM model, LSSVM can find the solution by solving a set of linear equations while classical SVM needs to solve a convex QP problem. As for the regression problem, given a training data set = $\{(x_1, y_1), (x_2, y_2), \ldots, (x_i, y_i)\}$, $x_i \in R^n$ and $y_i \in R$, and the separating hyper-plane in the feature space will be as Equation(1):

$$y(x) = w^T \varphi(x) \tag{1}$$

where w refers to the weight vector and $\varphi(x)$ is a nonlinear mapping from the input space to the feature space. Then the structural minimization is used to formulate the following optimization problem of the function estimation as Equation (2):

$$\min : \frac{1}{2} ||w||^2 + \frac{1}{2}\gamma \sum_{i=1}^{n} \varepsilon_i^2$$
$$\text{subject to} : y_i = w^T \varphi(x_i) + b + \varepsilon_i, \qquad i = 1, 2, \ldots, n \tag{2}$$

where γ refers to the regulation constant and ε_i to the error variable at time i, b to the bias term.

Define the Lagrange function as Equation (3):

$$L(w, b, \varepsilon, \alpha) = \frac{1}{2}||w||^2 + \frac{1}{2}\gamma \sum_{i=1}^{n} \varepsilon_i^2 - \sum_{i=1}^{n} \alpha_i \left\{ w^T \varphi(x_i) + b + \varepsilon_i - y_i \right\} \tag{3}$$

where α_i is the Lagrange multiplier.

Solving the partial differential of Lagrange function and introducing the kernel function, the final nonlinear function estimate of LSSVM with the kernel function can be written as Equation (4):

$$Y_i = f(X_i) = \sum_{i=1}^{n} \alpha_i K(X, X_i) + b \tag{4}$$

As for the selection of kernel function, this paper used the Gaussian radial basis function (RBF) as the kernel function, because RBF is the most effective for the nonlinear regression problems. And the RBF can be expressed as Equation (5):

$$K(X, X_i) = \exp\frac{-||X - X_i||^2}{2\sigma^2} \tag{5}$$

Through the above description, we can see that the selection of the regulation constant γ and Gaussian kernel function parameter σ has a significant influence on the learning effect and generalization ability of LSSVM. But the LSSVM model does not have a suitable method to select parameters, so we employ the global harmony search to realize the adaptive selection of parameters.

2.2. Global Harmony Search Algorithm in Parameters Determination of Least Squares Support Vector Machines Model

In music improvisation, musicians search for a perfect state of harmony by repeatedly adjusting the pitch of the instrument. Inspired by this phenomenon, HSA [36,37] is proposed by Geem *et al.* [36] as a new intelligent optimization search algorithm. However, every candidate solution in the fundamental HSA is independent to each other, which has no information sharing mechanism, thus, this characteristic also limits the algorithm efficiency. Lin and Li [38] have developed a GHSA which borrowed the concepts from swarm intelligence to enhance its performance [39]. The proposed GHSA procedure is illustrated as follows and the corresponding flowchart is shown in Figure 1.

Step 1: Define the objective function and initialize parameters.

Firstly, $f(x)$ is the objective function of the problem where x is a candidate solution consisting of N decision variables x_i and $L_{Bi} \leqslant x_i \leqslant U_{Bi}$. L_{Bi} and U_{Bi} are the lower and upper bounds for each variable. Besides, the parameters used in GHSA are also initialized in this step.

Step 2: Initialize the harmony memory.

The initialization process is done as:

- Randomly generate a harmony memory in the size of $2 \times HMS$ from a uniform distribution in the range$[L_{Bi}, U_{Bi}]$ $(i = 1, 2, 3, \ldots, n)$.
- Calculate the fitness of each candidate solution in the harmony memory and sort the results in ascending order.
- The harmony memory is generated by $[x_1, x_{2,}, \ldots, x_{HMS}]$.

Step 3: Improvisation.

The purpose of this step is to generate a new harmony. The new harmony vector $X' = \{x'_1, x'_2, \ldots, x'_n\}$ is generated based on the following rules:

Firstly, randomly generate r_1, r_2 in a uniform distribution of the range $[0, 1]$.

- If $r_1 <$ HMCR and $r_2 \geqslant$ PAR, then $x'_i = x'_i$. Palatino
- If $r_1 <$ HMCR and $r_2 <$ PAR, then $x'_i = \text{Rnd}\left(x_i^{gBest} - bw, x_i^{gBest} + bw\right)$, where bw is an arbitrary distance bandwidth (BW) and x_i^{gBest} is i^{th} dimension of the best candidate solution.
- If $r_1 \geqslant$ HMCR, then $x'_i = \text{Rnd}(L_{Bi}, U_{Bi})$.

Step 4: Update harmony memory.

If the fitness of the new harmony vector is better than that of the worst harmony, it will take the place of the worst harmony in the *HM*.

Step 5: Check the stopping criterion.

Terminate when the iteration is reached.

2.3. Fuzzy Time Series Generation

This paper proposes FCM model by using FCS algorithm with GA to process the raw data and to generate FTS. The flowchart is shown as Figure 1. Firstly, the number of clustering k is computed as the initial value. Secondly, the clustering center is obtained until the stop criteria of the algorithm are reached. Finally, the time series fuzzy membership is determined.

2.3.1. Fuzzy Time Series Model

A FTS is defined [40,41] as follows:

Definition 1: Let $Y(t)$ ($t = 0, 1, 2, \ldots$), a subset of real number, be the universe of the discourse on which fuzzy membership of f_i ($i = 1, 2, \ldots, n$) are defined. If $F(t)$ is a collection of $f_1, f_2, \ldots$, then $F(t)$ represents a FTS on $Y(t)$.

Definition 2: If $F(t)$ is caused by $F(t-1)$ only, the FTS relationship can be expressed as $F(t-1) \rightarrow F(t)$. Then let $F(t-1) = A_i$ and $F(t) = A_j$, so the relationship between $F(t-1)$ and $F(t)$ which is referred to as a fuzzy logical relationship can be denoted by $A_i \rightarrow A_j$.

We present the general definitions of FTS as follows:

Suppose U is divided into n subsets, such as $U = \{u_1, u_2, \ldots, u_n\}$. Then a fuzzy set A in the universe of the discourse of U can be expressed as Equation (6):

$$A = \frac{f_A(u_1)}{u_1} + \frac{f_A(u_2)}{u_1} + \ldots + \frac{f_A(u_n)}{u_n} \tag{6}$$

where $f_{A(u_i)}$ denotes the degree of membership of u_i in A with the condition of $f_{A(u_i)} \in [0,1]$.

2.3.2. Fuzzy C-Means Clustering Algorithm

Fuzzy c-means (FCM) [42] is a common clustering algorithm which could make one piece of data to cluster into multiple classes. Let $X = \{x_j | j = 1, 2, \ldots, n\}$ be the observation data set and $C = \{c_i | i = 1, 2, \ldots, k\}$ be the set of cluster centers. The results of fuzzy clustering can be expressed by membership function $U = \{u_{ij} | i = 1, 2, \ldots, k; j = 1, 2, \ldots, n\}$ where $u_{ij} \in [0,1]$ and u_{ij} is also limited by Equations (7) and (8).

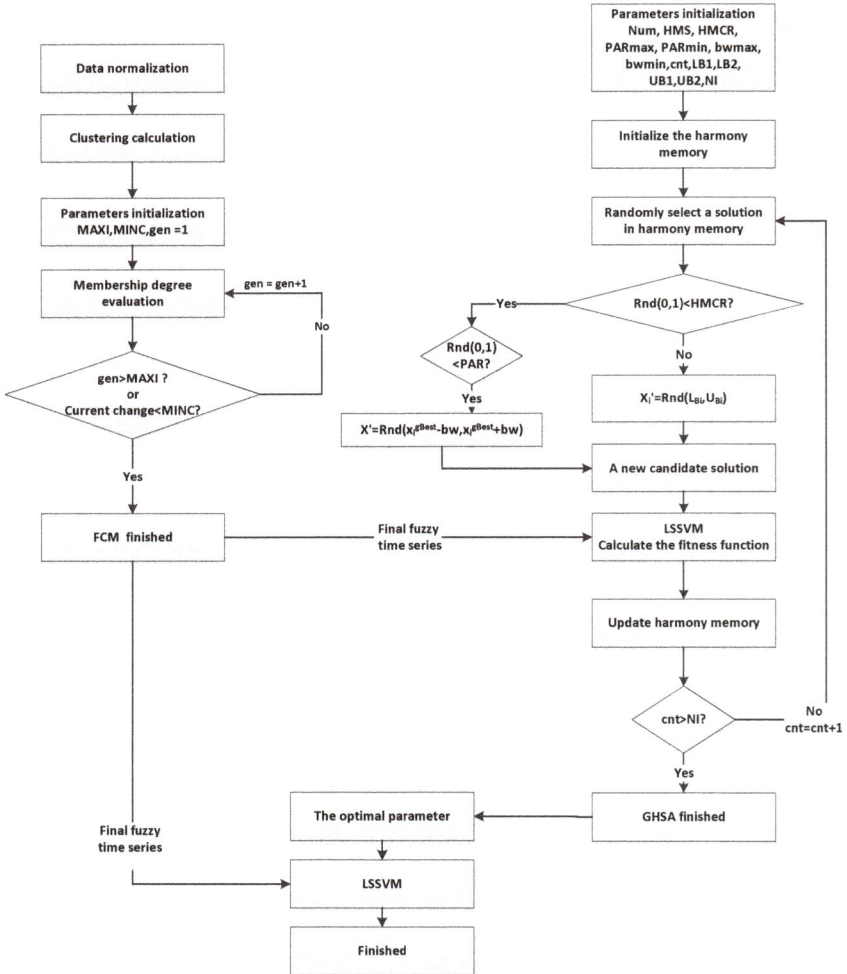

Figure 1. Global harmony search algorithm-fuzzy time series-least squares support vector machines (GHSA-FTS-LSSVM) algorithm flowchart.

$$\sum_{j=1}^{n} u_{ij} \in (0, n) \tag{7}$$

$$\sum_{i=1}^{k} u_{ij} = 1 \tag{8}$$

The objective function of FCM can be express as Equation (9):

$$J\left(u_{ij}, c_k\right) = \sum_{i=1}^{k} \sum_{j=1}^{n} \left(U\left(x_j, c_i\right)\right)^m \|x_j - c_i\|^2 \ (m > 1) \tag{9}$$

The cluster centers and the membership functions U are calculated by Equations (10) and (11):

$$c_i = \frac{\sum_{j=1}^{n} \left(u_{ij}\right)^m \cdot x_j}{\sum_{j=1}^{n} \left(u_{ij}\right)^m} \tag{10}$$

$$U\left(x_j, c_i\right) = u_{ij} = \frac{\left(\|x_j - c_i\|\right)^{-2/(m-1)}}{\sum_{l=1}^{k} \left(\|x_j - c_l\|\right)^{-2/(m-1)}} \tag{11}$$

where m is any real number named weight index, u_{ij} represents the membership of x_j in the ith cluster center and $\|x_j - c_i\|$ refers to the Euclidean distance between the real value x_j and the fuzzy cluster center c_i.

3. Numerical Example

3.1. Data Set

The experiment employs electric load data of Guangdong Province Industrial Development Database to compare the forecasting performances among the proposed GHSA-FTS-LSSVM model, GHSA-LSSVM model, GA-LSSVM, PSO-LSSVM and GHSA-LSSVM. The detailed data used in this paper is shown in Table 1. Among these data, the electric load data from January 2011 to December 2013 were used for model fitting and training, and the data from April to December 2014 were used to forecast.

Table 1. Monthly electric load in Guangdong Province from January 2011 to November 2014 (unit: thousand million W/h).

Date	Load	Date	Load	Date	Load
January 2011	284.1	May 2012	351.6	September 2013	372.3
February 2011	263.2	June 2012	353.1	October 2013	375.6
March 2011	339.8	July 2012	386.5	November 2013	386.4
April 2011	325.7	August 2012	376.1	December 2013	410.9
May 2011	336.2	September 2012	338	January 2014	384.5
June 2011	341	October 2012	343	February 2014	322.1
July 2011	371.7	November 2012	356.1	March 2014	389.2
August 2011	366.4	December 2012	362.4	April 2014	373.3
September 2011	329.8	January 2013	331	May 2014	387.6
October 2011	326.9	February 2013	278.1	June 2014	393.4
November 2011	331.4	March 2013	368.3	July 2014	429.8
December 2011	362.3	April 2013	357.2	August 2014	416.7
January 2012	341.5	May 2013	368.1	September 2014	379.9
February 2012	328.3	June 2013	373.3	October 2014	385.3
March 2012	358.7	July 2013	419.4	November 2014	398.2
April 2012	335.2	August 2013	426.6	December 2014	374.8

The procedure of data preprocessing is illustrated as follows:
Step 1: Data normalization

Before FCM, we normalized the original data by Equation (12):

$$X(i) = \frac{T(i) - T_{\min}}{T_{\max} - T_{\min}} \tag{12}$$

where $T(i)$ $(i = 1, 2, \ldots, n)$ is the set of time series which contains n observations, $T_{\min}$ and $T_{\max}$ refer to the minimum and maximum values of the data, $X(i)$ $(i = 1, 2, \ldots, n)$ is the normalized set of time series.

Step 2: Clustering calculation.

The number of clustering k is calculated by Equation (13) [43]:

$$k = \left[\frac{(T_{\max} - T_{\min}) \cdot (n-1)}{\sum_{t=2}^{n} |X(i) - X(i-1)|} \right] \tag{13}$$

where '[]' represents the rounded integer arithmetic. According to Equation (13), $k = 8$.

Step 3: Parameters initialization

We determine the maximum iteration $MAXI = 200$ and the minimum change of membership $MINC = 10^{-7}$. The performance of the algorithm depends on the initial cluster centers, so we need to specify a set of cluster centers at random.

Step 4: Update operator

If the objective function is better than the previous ones, the membership functions and cluster centers will be updated by Equations (10) and (11) after each iteration.

Step 5: Termination operator

In this paper, we use the iteration number and change of memberships as the termination operators. If the current iteration is larger than $MAXI$ or the current change of membership is smaller than $MINC$, the FCM finish its work and we can get the cluster centers.

After FCM, we got a set of clustering centers (set = {0.6115, 0.4595, 3949, 0.6668, 0.9491, 0.0732, 0.7485, 0.5416}), and the final time series fuzzy membership we got is shown in Table 2.

Table 2. The final fuzzy time series (FTS) (partly).

Date	FTS							
11 January	0.0104	0.0220	0.0338	0.0084	0.0036	0.9013	0.0063	0.0142
11 February	0.0128	0.0227	0.0307	0.0108	0.0053	0.8928	0.0085	0.0163
11 March	0.0000	1.0000	0.0000	0.0000	0.0000	0.0000	0.0000	0.0000
11 April	0.0064	0.0506	0.9181	0.0042	0.0011	0.0039	0.0026	0.0130
11 May	0.0115	0.7581	0.1842	0.0066	0.0013	0.0026	0.0036	0.0322
11 June	0.0026	0.9743	0.0106	0.0014	0.0002	0.0004	0.0007	0.0099
11 July	0.1255	0.0054	0.0030	0.8256	0.0022	0.0006	0.0210	0.0165
11 August	0.9547	0.0024	0.0012	0.0272	0.0006	0.0002	0.0037	0.0101
11 September	0.0005	0.0064	0.9911	0.0003	0.0001	0.0002	0.0002	0.0011
11 October	0.0029	0.0256	0.9604	0.0019	0.0005	0.0016	0.0011	0.0060
11 November	0.0046	0.0747	0.9032	0.0028	0.0006	0.0017	0.0016	0.0107
11 December	0.8419	0.0127	0.0058	0.0449	0.0019	0.0009	0.0098	0.0821

3.2. Global Harmony Search Algorithm-Least Squares Support Vector Machines Model

3.2.1. Parameters Selection by Global Harmony Search Algorithm

Before the GHSA we need to determine parameters. The parameters include the number of variables, the range of each variable [L_{Bi}, U_{Bi}] the harmony memory size (*HMS*), the harmony memory considering rate (*HMCR*), the value of *BW*, the pitch adjusting rate (*PAR*) and the number of iteration (*NI*).

In the experiments of GHSA, the larger harmony consideration rate (HMCR) is beneficial to the local convergence while the smaller HMCR can keep the diversity of the population. In this paper, we set the HMCR as 0.8. For the PAR, the smaller PAR can enhance the local search ability of algorithm while the larger PAR is easily to adjust search area around the harmony memory. In addition, the value of BW also has a certain impact on the searching results. For larger BW, it can avoid algorithm trapping into a local optimal and the smaller BW can search meticulously in the local area. In our experiment, we use a small PAR and a large BW in the early iterations of the algorithm, and with the increase of the NIs, BW is expected to be reduced while PAR ought to increase. Therefore, we adopt the following equations:

$$PAR = \frac{(PAR_{\max} - PAR_{\min}) * currentIteration}{NI} + PAR_{\min} \tag{14}$$

$$BW = BW_{\max} * \exp(\frac{\log(BW_{\min}/BW_{\max}) * currentIteration}{NI}) \tag{15}$$

In swarm intelligence algorithms, the global optimization ability of algorithm will be ameliorated by increasing the population size increase. However, the search time will also increase and the convergence speed will slow down as the population size becomes larger. On the contrary, if the population size is small, the algorithm will more easily be trapped in a local optimum. The original data consists of 48 sets. Combined with the relevant research experiences and a lot of experiments, we divided the number of data by the number of parameters and the quotient we got is 24, which is set to be the *HMS*. After continuous optimization experiments, we determined 20 as the *HMS* in GHSA.

In the LS-SVM model, the regularization parameter, γ, is a compromise to control the proportion of misclassification sample and the complexity of the model. It is used to adjust the empirical risk and confidence interval of data until the LS-SVM receiving excellent generalization performance. When the kernel parameter, σ, is approaching zero, the training sample can be correctly classified, however, it will suffer from over-fitting problem, and in the meanwhile, it will also reduce the generalization performance level of LS-SVM. Based on authors previous research experiences, the range of parameters γ and σ we determined in this paper are [0, 10000], [0, 100]. The parameters we select in GHSA are shown in Table 3.

Table 3. Parameters selection in GHSA.

Parameter	Value	Comment
num	$num = 2$	Number of variables
γ	$\gamma \in [0, 10000]$	Range of each variable
σ	$\sigma \in [0, 100]$	Range of each variable
HMS	$HMS = 20$	Harmony memory size
$HMCR$	$HMCR = 0.9$	HMS considering rate
PAR	$PAR_{\max} = 0.9, PAR_{\min} = 0.1$	Pitch adjusting rate
bw	$bw_{\max} = 1, bw_{\min} = 0.001$	Bandwidth
NI	$NI = 200$	Number of iteration

3.2.2. Fitness Function in Global Harmony Search Algorithm

Fitness function in GHSA is used to measure the fitness degree of generated harmony vector. Only if its fitness is better than that of the worst harmony in the harmony memory, it can replace the worst harmony. The fitness function is given Equation (16):

$$fit = 100 \times \frac{\sum_{i=1}^{n} \frac{|y_i - y_i'|}{y_i}}{n} \tag{16}$$

where n refers to the number of test sample, y_i refers to the observation value and y_i' to the predictive value in LSSVM.

Then we calculate the fitness function in GHSA by Equation (16). After finishing the GHSA, we have determined the optimal parameter $\gamma = 9746.7$ and $\sigma = 30.4$, then we establish LSSVM model to train historical data for forecasting the next electric load and get a set of output of LSSVM. At last, we denormalize the output of LSSVM.

3.2.3. Denormalization

After the GHSA we have determined the optimal parameter γ and σ, then we establish LSSVM model to train historical data for forecasting the next electric load. The outputs of LSSVM are normalized values, so we need to denormalize them to real values. Denormalization method is given by Equation (17):

$$v_{\text{real}} = v_i \times (max - min) + min \tag{17}$$

where *max* and *min* refers to the maximum and minimum value of the original data.

3.2.4. Defuzzification Mechanism

As indicated by several experiments that there are some inherent errors between actual values and fuzzy values. Therefore, it's necessary to estimate this kind of fuzzy effects to provide higher accurate forecasting performance. In this paper, we proposed an approach to adjust the fuzzy effects, namely defuzzification mechanism, as shown in Equation (18):

$$df_t = \text{AVG}\left(\frac{Y_{1t}}{FY_{1t}}, \frac{Y_{2t}}{FY_{2t}}, \ldots, \frac{Y_{it}}{FY_{it}}, \ldots, \frac{Y_{nt}}{FY_{nt}}\right) \tag{18}$$

where $t = 1, 2, \ldots 12$ for the twelve months in a year, n is the total year number of data set, $i = 1, 2, \ldots, n$ refers to the number of year and Y_{it}, FY_{it} is the actual value and fuzzy value of the ith year respectively. Thus, the final forecasting result can be expressed as Equation (19), and the defuzzification multipliers are shown in Table 4.

$$y_i' = y_i' * df_i \tag{19}$$

Table 4. Defuzzification multiplier of each month.

Month	Multiplier	Month	Multiplier
January	1.00244	July	1.00612
February	0.98222	August	1.00567
March	0.99931	September	1.00069
April	0.99522	October	0.99932
May	0.99772	November	1.00937
June	1.00493	December	0.99996

3.3. Performance Evaluation

We compare these proposals in different respects. First the proposed GHSA efficiency is compared with other optimization algorithms like HSA, PSO and GA. These appropriate algorithms are utilized to optimize the parameter γ and σ.

This experimental procedure is repeated 20 times for each optimization algorithm, and the performance comparison for different algorithms is represented in Figure 2, and the comparison of average fitness curves is presented in Table 5. We can see from Figure 2 and Table 5 that the values of the γ^{-1} obtained by the four algorithms are close to 0.0001, that is, all the search algorithms can achieve similar optimization, but the values of parameter, σ, as optimized by the different algorithms are not the same and this directly affects the fitness. The convergence speed of PSO is the fastest, however, due to the algorithm complexity, its running time is long. The execution time of HSA is the shortest, but the fitness is the worst. The running time of GHSA is equivalent to HSA, and the fitness of GHSA is optimal among four algorithms. In second group, the forecasting accuracy of the proposed algorithm is compared with ARIMA, GA-LSSVM [29], PSO-LSSVM [30] and the first group models. We take the mean absolute percentage error (MAPE), mean absolute error (MAE) and root mean squared error (RMSE) to evaluate the accuracy of the proposed method. The MAPE are shown in Equations (20)–(22):

$$MAPE = 100 \times \frac{\sum_{i=1}^{n} \frac{|y_i - y_i'|}{y_i}}{n} \tag{20}$$

$$MAE = \frac{\sum_{i=1}^{n} |y_i - y_i'|}{n} \tag{21}$$

$$RMSE = \sqrt{\frac{\sum_{i=1}^{n} (y_i - y_i')^2}{n}} \tag{22}$$

where n refers to the number of sample, y_i is the observation value and y_i' is the predictive value. According to the optimal value in Table 5, forecasting results of GHSA-FTS-LSSVM, HSA-LSSVM, GA-LSSVM and PSO-LSSVM models as shown in Table 6.

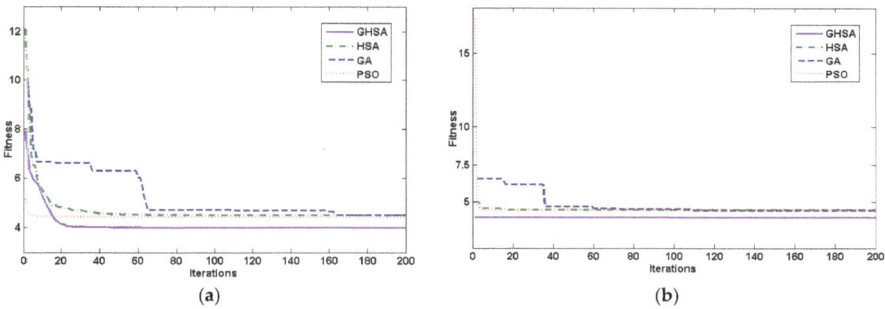

Figure 2. Comparison of (**a**) average fitness curves; and (**b**) best fitness curves.

Table 5. Performance comparison for different algorithms. Particle swarm optimization: PSO; harmony search algorithm: HAS; genetic algorithm: GA.

Algorithm	Fitness	γ^{-1}	σ	Running Time/s
GHSA	0.0397	0.00010	30.3977	9.2977
HSA	0.0489	0.00010	52.8422	8.2681
GA	0.0439	0.00010	52.8422	68.6248
PSO	0.0451	0.00011	22.3965	69.9352

Table 6. Forecasting results of GHSA-FTS-LSSVM, GHSA-LSSVM, GA-LSSVM, PSO-LSSVM and autoregressive integrated moving average (ARIMA) models (unit: thousand million W/h).

Time	Actual	GHSA-FTS-LSSVM	GHSA-LSSVM	GA-LSSVM [29]	PSO-LSSVM [30]	ARIMA
15 January	384.5	388.5989	387.094	387.066	393.205	399.142
15 February	352.1	379.4326	372.661	372.65	373.62	381.038
15 March	349.2	368.1298	355.006	355.01	352.864	359.864
15 April	373.3	359.5839	353.429	353.434	351.189	377.003
15 May	387.6	380.1802	366.55	366.545	366.026	362.173
15 June	393.4	392.6603	374.353	374.341	375.799	361.905
15 July	429.8	387.9569	377.522	377.506	379.962	399.488
15 August	416.7	395.6517	397.452	397.409	408.614	432.612
15 September	379.9	395.7048	390.271	390.239	397.814	423.027
15 October	385.3	376.4279	370.15	370.142	370.449	404.338
15 November	398.2	391.7981	373.098	373.086	374.179	390.129
15 December	374.8	380.8968	380.146	380.127	383.494	385.307
MAPE (%)	-	3.709	4.579	4.579	4.654	5.219
MAE	-	14.358	18.035	18.035	18.215	20.153
RMSE	-	18.180	21.914	21.921	21.525	23.0717

The excellent performance of the GHSA-FTS-LSSVM method is due to following reasons: first of all, we use FCM to process the original data, making the accurate load value become a set of input variables with fuzzy feature. Thus, the defects of the original data can be overcome and the implicit information is dug up. Secondly, the proposed algorithm employed the GHSA to improve the searching efficiency. Finally, LSSVM reduces the time of equation solving and improves the accuracy and generalization ability of the model.

4. Conclusions

Traditional electric load forecasting methods are based on the exact value of time series, but the electric power market is very complex, and the functional relations between variables are too difficult to describe, so this paper adopts the FTS model, and load values are defined as fuzzy sets. Then we compare the four algorithms GHSA, HSA, PSO and GA. According to the experimental results, it is obvious that GHSA which can find the optimal solution quickly and efficiently, is the best search algorithm in the LS-LSVM model. For the prediction accuracy, the MAPE of GHSA-FTS-LSSVM model is better than that of the GHSA-LSSVM which has no fuzzy processing. Also, our method has a better performance than the corresponding methods with GA and PSO.

Acknowledgments: This work was supported by Ministry of Science and Technology, Taiwan (MOST 104-2410-H-161-002) and Urban public utility government regulation Laboratory of Zhejiang university of Finance & Economics (No. Z0406413011/002/002).

Author Contributions: Yan Hong Chen and Wei-Chiang Hong conceived and designed the experiments; Wen Shen and Ning Ning Huang performed the experiments; Yan Hong Chen and Wei-Chiang Hong analyzed the data; Yan Hong Chen wrote the paper.

Conflicts of Interest: The authors declare no conflict of interest.

References

1. Ghayekhloo, M.; Menhaj, M.B.; Ghofrani, M. A hybrid short-term load forecasting with a new data preprocessing framework. *Electr. Power Syst. Res.* **2015**, *119*, 138–148. [CrossRef]
2. Kouhi, S.; Keynia, F. A new cascade NN based method to short-term load forecast in deregulated electricity market. *Energy Convers. Manag.* **2013**, *71*, 76–83. [CrossRef]
3. Chen, J.F.; Wang, W.M.; Huang, C.M. Analysis of an adaptive time-series autoregressive moving-average (ARMA) model for short-term load forecasting. *Electr. Power Syst. Res.* **1995**, *34*, 187–196. [CrossRef]
4. Nie, H.; Liu, G.; Liu, X.; Wang, Y. Hybrid of ARIMA and SVMs for short-term load forecasting. *Energy Proced.* **2012**, *16*, 1455–1460. [CrossRef]
5. Nedellec, R.; Cugliari, J.; Goude, Y. GEFCom2012: Electric load forecasting and backcasting with semi-parametric models. *Int. J. Forecast.* **2014**, *30*, 375–381. [CrossRef]
6. Kang, J.; Zhao, H. Application of improved grey model in long-term load forecasting of power engineering. *Syst. Eng. Proced.* **2012**, *3*, 85–91. [CrossRef]
7. Bahrami, S.; Hooshmand, R.A.; Parastegari, M. Short term electric load forecasting by wavelet transform and grey model improved by PSO (particle swarm optimization) algorithm. *Energy* **2014**, *72*, 434–442. [CrossRef]
8. Mandal, P.; Senjyu, T.; Urasaki, N.; Funabashi, T. A neural network based several-hour-ahead electric load forecasting using similar days approach. *Int. J. Electr. Power Energy Syst.* **2006**, *28*, 367–373. [CrossRef]
9. Ko, C.N.; Lee, C.M. Short-term load forecasting using SVR (support vector regression)-based radial basis function neural network with dual extended Kalman filter. *Energy* **2013**, *49*, 413–422. [CrossRef]
10. Yu, F.; Xu, X. A short-term load forecasting model of natural gas based on optimized genetic algorithm and improved BP neural network. *Appl. Energy* **2014**, *134*, 102–113. [CrossRef]
11. Pai, P.F. Hybrid ellipsoidal fuzzy systems in forecasting regional electricity loads. *Energy Convers. Manag.* **2006**, *47*, 2283–2289. [CrossRef]
12. Chandrashekara, A.S.; Ananthapadmanabha, T.; Kulkarni, A.D. A neuro-expert system for planning and load forecasting of distribution systems. *Int. J. Electr. Power Energy Syst.* **1999**, *21*, 309–314. [CrossRef]

13. Chen, Y.; Yang, Y.; Liu, C.; Li, L. A hybrid application algorithm based on the support vector machine and artificial intelligence: An example of electric load forecasting. *Appl. Math. Model.* **2015**, *39*, 2617–2632. [CrossRef]

14. Selakov, A.; Cvijetinović, D.; Milović, L.; Bekut, D. Hybrid PSO–SVM method for short-term load forecasting during periods with significant temperature variations in city of Burbank. *Appl. Soft Comput.* **2014**, *16*, 80–88. [CrossRef]

15. Hong, W.C. Electric load forecasting by seasonal recurrent SVR (support vector regression) with chaotic artificial bee colony algorithm. *Energy* **2011**, *36*, 5568–5578. [CrossRef]

16. Cortes, C.; Vapnik, V. Support-vector networks. *Mach. Learn.* **1995**, *20*, 273–297. [CrossRef]

17. Suykens, J.A.K.; Vandewalle, J. Least squares support vector machine classifiers. *Neural Proc. Lett.* **1999**, *9*, 293–300. [CrossRef]

18. Huang, S.C.; Wu, T.K. Integrating GA-based time-scale feature extractions with SVMs for stock index forecasting. *Expert Syst. Appl.* **2008**, *35*, 2080–2088. [CrossRef]

19. Huang, S.C. Integrating nonlinear graph based dimensionality reduction schemes with SVMs for credit rating forecasting. *Expert Syst. Appl.* **2009**, *36*, 7515–7518. [CrossRef]

20. Zhang, Y.; Lv, T. An empirical study on GPRS traffic forecasting based on chaos and SVM theory. *J. China Univ. Post. Telecommun.* **2010**, *17*, 41–44. [CrossRef]

21. Li, X.L.; Yi, Q.Z.; Liu, X.Y. Tax forecasting theory and model based on SVM optimized by PSO. *Expert Syst. Appl.* **2011**, *38*, 116–120.

22. Yang, Y.; Zhao, Y. Prevailing Wind Direction Forecasting for Natural Ventilation djustment in Greenhouses Based on LE-SVM. *Energy Proced.* **2012**, *16*, 252–258. [CrossRef]

23. Lee, W.J.; Hong, J. A hybrid dynamic and fuzzy time series model for mid-term power load forecasting. *Int. J. Electr. Power Energy Syst.* **2015**, *64*, 1057–1062. [CrossRef]

24. Efendi, R.; Ismail, Z.; Deris, M.M. A new linguistic out-sample approach of fuzzy time series for daily forecasting of Malaysian electricity load demand. *Appl. Soft Comput.* **2015**, *28*, 422–430. [CrossRef]

25. Pereira, C.M.; De Almeida, N.N.; Velloso, M.L.F. Fuzzy Modeling to Forecast an Electric Load Time Series. *Proced. Comput. Sci.* **2015**, *55*, 395–404. [CrossRef]

26. Sadaei, H.J.; Enayatifar, R.; Abdullah, A.H.; Gani, A. Short-term load forecasting using a hybrid model with a refined exponentially weighted fuzzy time series and an improved harmony search. *Int. J. Electr. Power Energy Syst.* **2014**, *62*, 118–129. [CrossRef]

27. Day, P.; Fabian, M.; Noble, D.; Ruwisch, G.; Spencer, R.; Stevenson, J.; Thoppay, R. Residential power load forecasting. *Proced. Comput. Sci.* **2014**, *28*, 457–464. [CrossRef]

28. Chen, S.M.; Kao, P.Y. TAIEX forecasting based on fuzzy time series, particle swarm optimization techniques and support vector machines. *Inf. Sci.* **2013**, *247*, 62–71. [CrossRef]

29. Mahjoob, M.J.; Abdollahzade, M.; Zarringhalam, R. GA based optimized LS-SVM forecasting of short term electricity price in competitive power markets. In Proceeding of the IEEE Conference on Industrial Electronics and Applications, Singapore, 3–5 June 2008; pp. 73–78.

30. Xiang, Y.; Jang, L. Water quality prediction using LS-SVM and particle swarm optimization. In Proceeding of the IEEE Second International Workshop on Knowledge Discovery and Data Mining, Moscow, Russia, 23–25 January 2009; pp. 900–904.

31. Mustaffa, Z.; Yusof, Y.; Kamaruddin, S.S. Gasoline Price Forecasting: An Application of LSSVM with Improved ABC. *Proced. Soc. Behav. Sci.* **2014**, *129*, 601–609. [CrossRef]

32. Zhai, M.Y. A new method for short-term load forecasting based on fractal interpretation and wavelet analysis. *Int. J. Electr. Power Energy Syst.* **2015**, *69*, 241–245. [CrossRef]

33. Zhang, W.Y.; Hong, W.C.; Dong, Y.; Tsai, G.; Sung, J.T.; Fan, G.F. Application of SVR with chaotic GASA algorithm in cyclic electric load forecasting. *Energy* **2012**, *45*, 850–858. [CrossRef]

34. Ghofrani, M.; Ghayekhloo, M.; Arabali, A.; Ghayekhloo, A. A hybrid short-term load forecasting with a new input selection framework. *Energy* **2015**, *81*, 777–786. [CrossRef]

35. Enayatifar, R.; Sadaei, H.J.; Abdullah, A.H.; Gani, A. Imperialist competitive algorithm combined with refined high-order weighted fuzzy time series (RHWFTS–ICA) for short term load forecasting. *Energy Convers. Manag.* **2013**, *76*, 1104–1116. [CrossRef]

36. Geem, Z.W.; Kim, J.H.; Loganathan, G.V. A new heuristic optimization algorithm: harmony search. *Simulation* **2001**, *76*, 60–68. [CrossRef]

37. Mahdavi, M.; Fesanghary, M.; Damangir, E. An improved harmony search algorithm for solving optimization problems. *Appl. Math. Comput.* **2007**, *188*, 1567–1579. [CrossRef]
38. Lin, J.; Li, X. Global harmony search optimization based ink preset for offset printing. *Chin. J. Sci. Instrum.* **2010**, *10*, 2248–2253. (In Chinese).
39. Wang, C.M.; Huang, Y.F. Self-adaptive harmony search algorithm for optimization. *Expert Syst. Appl.* **2010**, *37*, 2826–2837. [CrossRef]
40. Song, Q.; Chissom, B.S. Fuzzy time series and its models. *Fuzzy Sets Syst.* **1993**, *54*, 269–277. [CrossRef]
41. Egrioglu, E.; Aladag, C.H.; Yolcu, U.; Uslu, V.R.; Erilli, N.A. Fuzzy time series forecasting method based on Gustafson–Kessel fuzzy clustering. *Expert Syst. Appl.* **2011**, *38*, 10355–10357. [CrossRef]
42. Bezdek, J.C. *Pattern Recognition with Fuzzy Objective Function Algorithms*; Springer Science & Business Media: New York, NY, USA, 2013.
43. Yang, Y.W.; Lin, Y.P. Multi-step forecasting of stock markets based on fuzzy time series model. *Comput. Eng. Appl.* **2014**, *50*, 252–256.

energies

MDPI

Article

Research and Application of a Hybrid Forecasting Model Based on Data Decomposition for Electrical Load Forecasting

Yuqi Dong [1], Xuejiao Ma [2,*], Chenchen Ma [3] and Jianzhou Wang [2,3]

[1]   College of Law, Guangxi Normal University, Guilin 541004, China; 18074845560@163.com
[2]   School of Statistics, Dongbei University of Finance and Economics, Dalian 116023, China;
      wangjz@dufe.edu.cn
[3]   School of Mathematics and Statistics, Lanzhou University, Lanzhou 730000, China; chenchenma133@163.com
*   Correspondence: xuejiaomadufe@163.com; Tel.: +86-186-2443-8689

Academic Editor: Wei-Chiang Hong
Received: 26 October 2016; Accepted: 2 December 2016; Published: 14 December 2016

Abstract: Accurate short-term electrical load forecasting plays a pivotal role in the national economy and people's livelihood through providing effective future plans and ensuring a reliable supply of sustainable electricity. Although considerable work has been done to select suitable models and optimize the model parameters to forecast the short-term electrical load, few models are built based on the characteristics of time series, which will have a great impact on the forecasting accuracy. For that reason, this paper proposes a hybrid model based on data decomposition considering periodicity, trend and randomness of the original electrical load time series data. Through preprocessing and analyzing the original time series, the generalized regression neural network optimized by genetic algorithm is used to forecast the short-term electrical load. The experimental results demonstrate that the proposed hybrid model can not only achieve a good fitting ability, but it can also approximate the actual values when dealing with non-linear time series data with periodicity, trend and randomness.

Keywords: electrical load forecasting; data decomposition; genetic algorithm; generalized regression neural network

1. Introduction

The electric power industry plays a pivotal role in the national security, social stability and all aspects of people's life. As is known to all, electricity, as one of the most important energy resources, is difficult to store. A great variety of instability factors can affect the electric system, such as emergencies, holidays, population changes, the weather and more [1]. Therefore, there is a high demand for the generation, transmission and sales of electricity, because excess supply can result in wasted energy resources and in case of excess demand the need for electricity cannot be satisfied. Therefore, performing load forecasting based on the historical data has been a basic task in the operation of electric systems [2]. With the rapid development of society and continuous improvement of economic levels, people have gradually shown a higher desire for electricity, which poses a huge challenge to the forecasting accuracy. A higher accuracy can improve the electric energy usage, enhance the safety and reliability of power grid and have a big impact on all sections in the electric power system. Accurate forecasting of electrical load plays a significant role, which can be reflected in the following aspects:

◆   Improve the social and economic benefits. The electrical power sector is supposed to ensure a good social benefit through providing safe and reliable electricity and improving the economic benefits considering the cost problems. Thus, the electrical load forecasting is beneficial for electrical power system to achieve the economic rationality of power dispatching.

◆ Ensure the reliability of electricity supply. Whether the power generation or supply, equipment needs periodical overhauls to ensure the safety and reliability of electricity. However, when to overhaul or replace the equipment should be based on accurate electrical load forecasting results.

◆ Plan for electrical power construction. The construction of electrical power production sites cannot stay unchanged, and should be adjusted and perfected, to satisfy the demands of a constantly changing future with the progress of society and development of the economy.

There are a great number of methods to forecast the electrical load, and in general the electrical load forecasting can be divided into three types, according to the applied field and forecasting time:

◆ Long-term electrical load forecasting. This means a time interval above five years and is usually conducted during the planning and building stage of the electrical system, which considers the characteristics of the electrical system and the development tendencies of the national economy and society;

◆ Middle-term electrical load forecasting. It is mainly applied in the operation stage of the electrical power system, for direction of the scientific dispatch of power, arrangement of overhauling and so on;

◆ Short-term electrical load forecasting. It plays a pivotal role in the whole electrical system and is the most important part, for it is the basis of long- and middle-term electrical load forecasting. Besides, it can ensure the stable and safe operation of the electrical power system based on the forecasting data.

Electrical load forecasting is a very complicated work. On the one hand, the electrical power system itself is complex and of large size. On the other hand, the electrical market closely combines the electrical power system with the whole society. Therefore, to properly monitor changes of the electrical load has become increasingly crucial for utilities so as to secure a steady power supply and make a suitable plans for investing in power facilities [3]. On the contrary, the inaccurate electrical load forecasting would be counterproductive. The overestimated future electrical load will result in an unnecessary generation of electrical power; while the underestimated forecasting would lead to trouble in offering sufficient electrical power, resulting in high losses for per peaking unit [4,5]. In addition, the inaccurate electrical load forecasting would also directly increase the operating costs. Therefore, to develop a better forecasting method and improve the forecasting ability has been more and more imperative, which is a both significant and challenging task [6].

In recent years, the study of short-term electrical load time series forecasting has mainly included four aspects, which are classic forecasting methods, modern forecasting methods, combined forecasting methods and hybrid forecasting methods [7].

The classic forecasting models refer to regression analysis, time series analysis and so on. The regression analysis models regard the influencing factors of time series as independent variables, and the historical data as the dependent variable, ensuring the relationship between the series and influencing factors. These methods are based on the analysis of historical data, so they can better model the history, however, as time goes by, the forecasting effect of regression analysis models will become weaker and weaker. The regression analysis process is easy, and the parameter estimation methods are complete; however, when dealing with non-linear time series data, the forecasting quality is bad and the forecasting accuracy is low. Another drawback is that it is difficult to select the influencing factors owing to the complexity of the objective data [8]. Time series forecasting aims to construct mathematical models based on the statistics of historical data, and it requires relatively small datasets and achieves a fast analysis speed, which can capture the variation trends of the recent data. However, it has a high requirement for stability, so when the influence of random factors is strong, the model will achieve a bad forecasting effect and low forecasting accuracy.

The modern forecasting methods include artificial intelligence neural networks [9,10], chaotic time series methods [11], expert system forecasting methods [12], grey models [13,14], support vector machines [15,16], fuzzy systems [17], self-adaptable models [18], optimization algorithms and so on. The artificial neural networks (ANNs) can simulate the human brain to realize the intelligent dealing, and it can obtain a good forecasting performance when addressing the non-structural and non-linear time series data owing to their ability of self-adaptability, self-learning and memory. In 1991, Park [19] first applied ANNs in electrical load forecasting, proving the good performance of the model and at the same time concluding that ANNs were applicable in electrical load forecasting. Since then a large number of researchers have utilized many types of ANNs to forecast the time series [20–22]; however, ANNs also have its own limitations and disadvantages: (1) It is difficult to determine scientifically the number of layers and neurons of a network structure; (2) ANNs have a relatively slow self-learning convergence rate, which makes it easy to fall into a local minimum; (3) The ability to express the fuzzy awareness of human brain is not strong. Therefore, other methods, such as support vector machine (SVM) and evolution algorithms (EA), are used to overcome the dependence of ANNs on the samples, enhance the extrapolation power, and reduce the learning time. Pandian [23] and Pai [24] applied ANNs in electrical load forecasting systems. The optimization algorithms are enlightened by the biological evolution, which is effective in dealing with complicated problems. Optimization algorithms are usually combined with other forecasting methods, with the aim of selecting and recognizing parameters. For example, in the aspect of ANNs, optimization algorithms do not depend on subjective experience to determine parameters; instead, it can select more reasonable parameters through objective algorithms.

In view of the limitations and accuracy errors of single algorithms, they cannot be adapted to all situations; therefore, the combined models have gradually become the development tendency currently [25]. The combined forecasting models were initially proposed by Bates and Granger who proved that the linear combination of two forecasting models could obtain better forecasting results than the single models alone. Xiao et al. [26] and Wang et al. [27] also proved that the forecasting accuracy of the combined model were higher than that of a single model. The basic principles of the combined forecasting methods are to integrate the forecasting output results of different single models based on certain weights, narrowing the value range of the forecasting down to a smaller scale. A problem is supposed to be studied from different angles instead of a single angle, and this is why the combined forecasting model is needed. The information obtained from each single forecasting method is not the same, and a weight is necessary to express the outputs of each single model more comprehensively in order to retain the original valuable information. Recently the combined forecasting models have been commonly used to solve forecasting issues, but how to select the single model properly and distribute the weight reasonably is a challenging task.

The theory of hybrid algorithms can get over the shortcomings of the single forecasting model through integrating two or more than two single models. As discussed above, the single models have their own advantages and disadvantages when dealing with different forecasting problems. In comparison, the hybrid forecasting methods can increase the forecasting accuracy through determining an optimal combination and putting the advantages of single models into full play. In other words, the hybrid algorithms can integrate many different forecasting techniques to solve practical problems in practice. For example, the blind number theory can be applied in middle- and long-term electrical load forecasting to build a hybrid model, which can enhance the forecasting effects well due to the irregular nature of electrical load time series.

Affected by many factors, the complexity of time series continues improving, and several techniques are utilized to solve the forecasting problems of time series. Azimi et al. [28] built a novel hybrid model to forecast the short-term electrical load, because a single model cannot figure out the characteristics of the time series data. Khashei and Bijari [29] considered that there was no a single model that could ensure the real process of the data generation. Shukur and Lee [30] proposed a hybrid model, including ANN and auto regressive integrated moving average (ARIMA), taking full advantage of the linear and non-linear advantages of the two models. Considerable experimental results demonstrate that the forecasting accuracy of the hybrid model represents a great improvement when compared with other single models. Aiming to improve the forecasting quality, Niu [31] built a new hybrid ANN model and combined some statistical methods to conduct forecasting. Lu and Wang [32] developed a growing hierarchical self-organizing map (SOM) with support vector machine (SVM) to forecast the product demand. Okumus and Dinler [33] integrated the adaptive neuro-fuzzy inference system and ANNs to forecast the wind power and their experimental results proved that the proposed hybrid model was better than applying the single model. Che and Wang [34] put forward the SVMARIMA hybrid model with SVM and ARIMA to forecast both the linear and non-linear trends more accurately. Meng et al. [35] developed a hybrid model for short-term wind speed forecasting by applying wavelet packet decomposition, crisscross optimization algorithm and artificial neural networks, and their experimental results showed that the proposed hybrid model had the minimum mean absolute percentage error, regardless of whether one-step, three-step or five-step prediction was used. Elvira [36] selected five forecasting methods to forecast the electrical load in summer and winter in the southeastern region of Oklahoma respectively. The empirical results showed that there was no one model that could always perform the best in all conditions, and differences in the original time series data and the evaluation metrics used to measure errors would both have an impact on the selection of the optimal model. Wu et al. [37] proposed a hybrid forecasting method based on seasonal index adjustment, and applied it in the forecasting of short-term wind speed and electrical load. The experimental results indicated that compared with the method without seasonal index adjustment, the proposed hybrid model could achieve a better forecasting result.

As discussed above, the single modela cannot satisfy the requirementa for forecasting accuracy in practice, and there is no one model applicable in any situation. Given that the actual data will be affected by various factors, which are difficult to recognize and measure, and it is not possible to take every related factor into consideration, the model is supposed to be built based on some key factors that can be extracted. The establishment of the hybrid model has become the mainstream currently. Therefore, this paper proposes a hybrid forecasting model considering periodicity, trend and randomness for electrical load time series. The contributions of the model are summarized as follows:

(1)   The time series data have the characteristics of continuity, periodicity, trend and randomness, and considerable work has been done to select suitable models and the optimize the model parameters; however, few studies focus on building forecasting models based on the characteristics of the time series data. Therefore, the initial contribution of this paper is to decompose the time series data. Based on the traditional additive model, the layer-upon-layer decomposition and reconstitution method is applied to improve the forecasting accuracy. Then according to the data features after decomposition, suitable models could be found to perform the forecasting. Through effective decomposition of the data and selection of reasonable model, the forecasting quality and accuracy could be improved to a great degree.

(2)   This paper uses the generalized regression neural network (GRNN) to improve the forecasting performance. The data after decomposition have noises, so the empirical mode decomposition (EMD) is applied to reduce the noise in the data. Then the genetic algorithm (GA) is utilized to optimize the GRNN to conduct the forecasting to enhance the forecasting accuracy of the single model.

(3)   The practical application of the proposed hybrid model in this paper is to forecast the short-term electrical load in New South Wales of Australia, and compare it with the single models and models without decomposition. The forecasting results demonstrate that the proposed model has a strong non-linear fitting ability and good forecasting quality for electrical load time series. Both the simulation results and the forecasting process could fully show that the hybrid model based on the data decomposition has the features of small errors and fast speed. The algorithm applied in the electrical power system is not only applicable, but also effective.

The rest of this paper is organized as follows: Section 2 describes the method and Section 3 introduces the detailed steps of the hybrid model, respectively. The experimental results are shown in Section 4. Section 5 presents the conclusions.

2. Methods

Conducting an accurate electrical load forecasting needs better developed forecasting methods and it is imperative to have improved forecasting abilities. This paper proposes a hybrid model to perform short-term electrical load forecasting, and this part introduces the fundamental methods, including additive model of time series, moving average model, cycle adjustment model, empirical mode decomposition and generalized regression neural network.

2.1. Additive and Multiplicative Model of Time Series

In general, a time series can be decomposed into two types of models through data transformation, including the additive model and the multiplicative model, as shown in Equations (1) and (2):

$$Y_t = S_t + T_t + C_t + R_t \tag{1}$$

$$Y_t = S_t \times T_t \times C_t \times R_t \tag{2}$$

where S_t is a seasonal item, indicating the law of transformation of time series with the season, which exists objectively. Actually, the electrical load time series always shows a seasonal cycle fluctuation; that is to say, the sequence will change repeatedly and continuously with time, showing a periodicity rule. Therefore, this paper classifies the seasonal item into a periodic item considering the clarity of expression. T_t is a trend item, denoting the law of transformation of time series with the trend. It mainly represents a long-term changing rule, because the time series will keep increasing, decreasing or remain stable. C_t is a periodic item and it indicates a periodic and non-seasonal law of transformation of time series with time. The number of a cycle fluctuation periods is expressed as h. R_t is a random item, which indicates the random change. Through decomposition, the original time series could be transformed into a stationary time series, which could achieve a good fitting and forecasting result.

2.2. Moving Average Model

The original time series will show the features of continuity, periodicity, trend and randomness. In order to eliminate the features and obtain a smoother time series, the moving average model will be applied. The algorithm principle is to calculate the average of the historical data, and the average is regarded as the next forecasting value until the final forecasting goal is realized. In other words, a new value will replace the old value, among which the number of items of the moving average is fixed. The detailed calculation equation is described as follows:

$$M_t^{(1)} = \frac{y_t + y_{t-1} + \cdots y_{t-N-1}}{N} = M_{t-1}^{(1)} + \frac{y_t - y_{t-N}}{N}, t \geq N \tag{3}$$

where $X = \{y_1, y_2, \cdots y_t\}$ is the original time series, N is the number of average, $M_t^{(1)}$ is the moving average in the t-th period, y_t is the observed value in the t-th period and N is the number of fixed items. The forecasting equation is:

$$\hat{y}_{t+1} = M_t^{(1)} \tag{4}$$

2.3. Periodic Adjustment Model

The essence of the cycle adjustment is to summarize the cycle variation law based on the periodic historical data. Assume that a group of periodic data $\{c_t, t \in \{1, 2, \cdots T\}\}$, it is divided into l groups and the number of data in each group is h $(T = l \times h)$. The data series can be defined as:

Definition 1. *The time series data $\{c_t, t \in \{1, 2, \cdots T\}\}$ is decomposed into $\{c_{11}, c_{12}, \ldots, c_{1s}, \ldots, c_{1h}\}$, $\{c_{21}, c_{22}, \ldots, c_{2s}, \ldots, c_{2h}\}, \{c_{k1}, c_{k2}, \ldots, c_{ks}, \ldots, c_{km}\}, \ldots, \{c_{l1}, c_{l2}, \ldots, c_{ls}, \ldots, c_{lh}\}$ $(k = 1, 2, \ldots, l; s = 1, 2, \ldots, h)$. c_{ks} means s-th data in the k-th period.*

The average of each group can be used to approximate the periodic average [38]. The s-th average period is:

$$\bar{c}_s = (c_{1s} + c_{2s} + \cdots c_{ls})/l \ (s = 1, 2, \ldots h) \tag{5}$$

The average of all data is:

$$Z = (\bar{c}_1 + \bar{c}_2 + \cdots \bar{c}_h)/h \tag{6}$$

The periodic value after adjustment is:

$$\hat{c}_s = \bar{c}_s - Z(s = 1, 2, \ldots, h) \tag{7}$$

Equations (5)–(7) represent the periodic variation law.

2.4. Empirical Mode Decomposition

The empirical mode decomposition, initially proposed in 1998, belongs to the data mining methods, which play a crucial role in dealing with the non-linear data Currently, it has been applied in many fields, such as geography [39], economics [40] and so on. EMD is a type of new method to divide the same non-stationary into different frequencies. The sequence of the composed different signal scales is called intrinsic mode function (IMF), which is the non-linear and stationary signal. IMF has an obvious feature that the wave amplitude changes with time. For given signal $x(t) \in R^t$, the detailed steps of EMD are described as follows (as shown in Figure 1I):

Step 1. Find all the local extreme points of $x(t)$.

Step 2. For all local extreme points of $x(t)$, build the envelope function of the signal, respectively, which can be denoted as $e_{max}(t)$ and $e_{min}(t)$.

Step 3. Calculate the average of the envelope function:

$$e_m(t) = \frac{e_{min}(t) + e_{max}(t)}{2} \tag{8}$$

Step 4. Calculate the differential function between signal $x(t)$ and the envelope average function

$$h(t) = x(t) - e_m(t) \tag{9}$$

Step 5. Replace the original signal $x(t)$ with $h(t)$, and repeat above steps from Step 2 to Step 4 until all averages of envelope function tends to zero. In this way an IMF $c_1(t)$ is decomposed.

Step 6. $c_1(t)$ represents the component with the highest frequency, so the low frequency of the original signal is $r_1(t)$:

$$r_1(t) = x(t) - c_1(t) \tag{10}$$

$$r_2(t) = r_1(t) - c_2(t) \tag{11}$$

$$r_n(t) = r_{n-1}(t) - c_n(t) \tag{12}$$

Step 7. For $x_1(t)$, repeat Step 2, Step 3 and Step 4, and the second IMF $c_2(t)$ can be obtained until the differential function $r_n(t)$ is a constant function or monotone function. Finally, the original signal $x(t)$ can be represented by IMF $c_j(t), j = 1, 2, \cdots, n$ and $r_n(t)$ as shown in Equation (13):

$$x(t) = \sum_{j=1}^{n} c_j(t) + r_n(t) \tag{13}$$

The EMD steps of the time series are shown in Figure 1I, and the pseudo code of EMD is described in Algorithm 1 below.

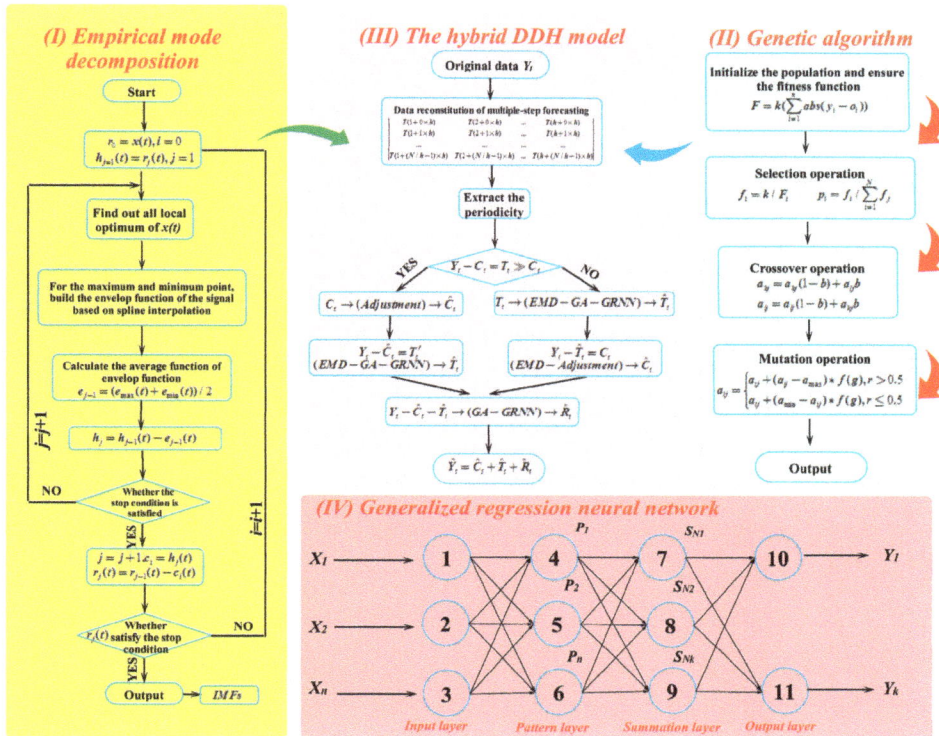

Figure 1. Steps of the main methods and proposed hybrid model in this paper.

Algorithm 1: Pseudo code of Empirical Mode Decomposition

Input: $x_s^{(0)} = (x^{(0)}(1), x^{(0)}(2), \ldots, x^{(0)}(n))$—a sequence of sample data.

Output: $\hat{x}_s^{(0)} = (\hat{x}^{(0)}(l+1), \hat{x}^{(0)}(l+2), \ldots, \hat{x}^{(0)}(l+n))$—a sequence of denoising data.

Parameters:

   δ—represent a random number in the algorithm with the value between 0.2 and 0.3.

   T—a parameter describing the length of the original *electrical load* time series data.

1: /***Initialize** residue $r_0(t) = x(t), i = 1, j = 0$; Extract local maxima and minima of $r_{i-1}(t)$.*/

2: FOR EACH $(j = j + 1)$ DO

3: FOR EACH $(i = 1 : n)$ DO

4: WHILE (Stopping Criterion $SD_j = \sum\limits_{t=0}^{T} \dfrac{|h_{i,j-1}(t) - h_{i,j}(t)|^2}{[h_{i,j-1}(t)]^2} > \delta$) DO

5: Calculate the upper envelope $U_i(t)$ and $L_i(t)$ via cubic spline interpolation.

6: $m_i(t) = \dfrac{U_i(t) + L_i(t)}{2}$ /* Mean envelope */; $h_i(t) = r_{i-1}(t) - m_i(t)$/* ith component */

7: /*Let $h_{i,j}(t) = h_i(t)$, with $m_{i,j}(t)$ being the mean envelope of $h_{i,j}(t)$*/

8: END WHILE

9: Calculate $h_{i,j}(t) = h_{i,j-1}(t) - m_{i,j-1}(t)$

10: /*Let the jth IMF be $IMF_i(t) = h_{i,j}(t)$; Update the residue $r_i(t) = r_{i-1}(t) - IMF_i(t)$*/

11: END DO

12: END DO

13: Return $x(t) = \sum\limits_{j=1}^{n} c_j(t) + r_n(t)$ /* The noise reduction process is finished */

2.5. Generalized Regression Neural Network (GRNN)

The generalized regression neural network, first proposed by Specht in 1991, is a type of radial basis function neural network (RBF). The theory of GRNN is based on non-linear regression analysis, and in essence, the purpose of GRNN is to calculate y with the biggest probability value based on the regression analysis of dependent variable Y and independent variable x. Assume that joint probability density function of the random variable x and y is $f(x, y)$, and the observed value x is known as X, so the regression of y about x is:

$$\hat{Y} = E(y/X) = \frac{\int_{-\infty}^{\infty} y f(X, y) dy}{\int_{-\infty}^{\infty} f(X, y) dy} \tag{14}$$

The density function $f(X, y)$ can be estimated from the sample data set $\{x_i, y_i\}_{i=1}^n$ by applying Parzen non-parametric estimation:

$$\hat{f}(X, y) = \frac{1}{n(2\pi)^{\frac{p+1}{2}} \sigma^{p+1}} \sum_{i=1}^n \exp\left[-\frac{(X - X_i)^T (X - X_i)}{2\sigma^2}\right] \exp\left[-\frac{(X - Y_i)^2}{2\sigma^2}\right] \tag{15}$$

where X_i and Y_i is the sample observed value of x and y, n is the sample size, p is the number of dimension of random variable x and σ is the smoothing factor. $\hat{f}(X, y)$ can replace $f(X, y)$ of Equation (15), so the function after transformation is:

$$\hat{Y}(X) = \frac{\sum_{i=1}^n \exp\left[-\frac{(X - X_i)^T (X - X_i)}{2\sigma^2}\right] \int_{-\infty}^{\infty} y \exp\left[-\frac{(Y - Y_i)^2}{2\sigma^2}\right] dy}{n \sum_{i=1}^n \exp\left[-\frac{(X - X_i)^T (X - X_i)}{2\sigma^2}\right] \int_{-\infty}^{\infty} \exp\left[-\frac{(Y - Y_i)^2}{2\sigma^2}\right] dy} \tag{16}$$

For $\int_{-\infty}^{\infty} z e^{-z^2} dz = 0$, after calculating the two integration, the output of GRNN can be $\hat{Y}(X)$ obtained as follows:

$$\hat{Y}(X) = \frac{\sum_{i=1}^n y \exp\left[-\frac{(X - X_i)^T (X - X_i)}{2\sigma^2}\right]}{\sum_{i=1}^n \exp\left[-\frac{(X - X_i)^T (X - X_i)}{2\sigma^2}\right]} \tag{17}$$

After obtaining the training samples of GRNN, the training process of the network involves optimizing the smoothing parameter σ. In order to improve the fitting ability of GRNN, σ needs to be optimized, which indicates the importance of optimizing the smoothing parameter σ in GRNN.

As for the structure of GRNN, it is similar to that of RBF, including input layer, pattern layer, summation layer and output layer. The corresponding network input is $X = [x_1, x_2, \ldots, x_n]$, and its output is $Y = [y_1, y_2, \ldots, y_n]^T$, which are described below.

(1) *Input layer*

The number of neuron of the input layer is the same as the dimension number of input variable, which plays a role in transferring signals.

(2) *Pattern layer*

The number of neuron of the pattern layer is the same as the number of learning samples, and the transfer function is

$$P_i = \exp\left[-\frac{(X - X_i)^T (X - X_i)}{2\sigma^2}\right], i = 1, 2, \ldots, n \tag{18}$$

where X is the input variable of the network, and X_i is the learning sample of ith neuron.

(3) Summation layer

Two methods can be applied to calculate the neuron. One is shown in Equation (10):

$$\sum_{i=1}^{n} \exp\left[-\frac{(X - X_i)^T (X - X_i)}{2\sigma^2}\right] \tag{19}$$

where the arithmetic sum of each neuron is calculated, the link weight is 1, and the transfer function is:

$$S_D = \sum_{i=1}^{n} P_i \tag{20}$$

The other method is:

$$\sum_{i=1}^{n} Y_i \exp\left[-\frac{(X - X_i)^T (X - X_i)}{2\sigma^2}\right] \tag{21}$$

where the weighted arithmetic sum of each neuron is calculated, and the link weight between the i-th neuron and j-th molecular sum neurons is the j-th element of i-th output sample Y_j. The transfer function is:

$$S_{Nj} = \sum_{i=1}^{n} y_{ij} P_{ij}, j = 1, 2, \ldots, k \tag{22}$$

(4) Output layer

The number of neuron of output layer is the same as the dimension number k of output variable. The output of summation layer is divided by each neuron as shown in Equation (23):

$$y_j = \frac{S_{Nj}}{S_D} \tag{23}$$

Then there are some weights in GRNN to connect different layers, and the least mean squares and differential chain rule are applied to adjust them. Initially, we define the least mean square of each neuron in the output layer:

$$E_k = [d_k(X) - F_k(W, X)]^2 / 2, k = 1, 2, \ldots, K \tag{24}$$

where $d_k(X)$ is the expected output, $F_k(W, X)$ is the actual output. E_k can arrive at the smallest value through adjusting the weights according to Equation (25) by using the least mean squares method:

$$\Delta w_{ki}(n) = \eta_k \left(-\frac{\partial E_k}{\partial w_{ki}}\right), i = 1, 2, \ldots, M; k = 1, 2, \ldots, K \tag{25}$$

where η_k is the learning rate. Therefore, the key to realizing the least square mean is to solve $(-\partial E_k / \partial w_{ki})$, so by using the differential chain rule, we can get:

$$-\frac{\partial E_k}{\partial w_{ki}} = -\frac{\partial E_k}{\partial F_k(W, X)} \frac{\partial F_k(W, X)}{\partial w_{ki}} \tag{26}$$

where $-\partial E_k / \partial F_k(W, X) = d_k(X) - F_k(W, X)$, which can be denoted as δ_k. Then we can get $(-\partial E_k / \partial w_{ki}) = \delta_k y_i$ according to Equation (27):

$$\frac{\partial F_k(W, X)}{\partial w_{ki}} = \frac{\partial}{\partial w_{ki}} \left(\sum_{i=1}^{M} w_{ki} y_{ki}\right) = y_i \tag{27}$$

so $\Delta w_{ki}(n) = \eta_k \delta_k y_i$, where y_i is the output of i-th neuron in the hidden layer, and the input of kth neuron in the output layer. The detailed structure of GRNN is described in Figure IV.

3. The Proposed Hybrid Model

In the proposed data decomposition hybrid model (DDH), we initially remove the periodicity in the original series, and then the EMD-GA-GRNN is applied to forecast the electrical load time series without periodicity. After that the periodicity is added to the forecasted time series by using the additive model. This part will introduce the basic ideas of both DDH and EMD-GA-GRNN.

3.1. Genetic Algorithm

The genetic algorithm is based on the natural selection rule and biological evolution principle, and its basic idea is to generate a set of initial solutions (population) in the problem space. Each group of solutions is regarded as the individuals in the population, which is defined as a chromosome. In the searching process, the adaptive value of chromosomes is the standard used to evaluate and select individuals. In the next generation, new individuals are generated through crossover and mutation operations, becoming a new generation of the population [41]. The above steps are repeated so that the chromosome can converge to a desired optimum value and solution. GA is applied in this paper to optimize GRNN, and the detailed steps are described as follows (as shown in the pseudo code of Algorithm 2 and Figure 1II):

Step 1. Initialize the population. Each individual in the population is a real number, with a known net structure, the initial values can form a neural network with structure, weight value and threshold value.

Step 2. Ensure the fitness function. The fitness value F is the absolute error values between the forecasting output and expected output calculated by Equation (28):

$$F = k(\sum_{i=1}^{n} abs(y_i - o_i)) \tag{28}$$

where n is the number of the output node of the network, y_i is the expected output of ith node, o_i is the forecasting output of ith node, and k is the coefficient.

Step 3. Selection operation. This operation is based on the proportion of the fitness, and the selection probability of each individual i is p:

$$f_i = \frac{k}{F_i} \tag{29}$$

$$p_i = \frac{f_i}{\sum_{i=1}^{n} f} \tag{30}$$

where F_i is the fitness of individual i, and the smaller fitness is better. Before the selection operation, the reciprocal of fitness should be calculated. k is the coefficient and N is the number of individual in the population.

Step 4. Crossover operation. The individual is coded by using the real number, and the crossover operation in the jth position between kth chromosome a_k and a_l lth chromosome a₁:

$$a_{kj} = a_{kj}(1 - b) + a_{lj}b \tag{31}$$

$$a_{lj} = a_{lj}(1 - b) + a_{kj}b \tag{32}$$

where b is a random number of [0,1].

Step 5. Mutation operation. Select the *j*-th gene of *i*-th individual to conduct the mutation operation, and the method is:

$$a_{ij} = \begin{cases} a_{kj} = a_{kj}(1-b) + a_{lj}b \\ a_{ij} + (a_{ij} - a_{max}) * f(g), r > 0.5 \\ a_{ij} + (a_{min} - a_{ij}) * f(g), r \le 0.5 \end{cases} \tag{33}$$

where a_{max} is the upper bound of gene a_{ij}, a_{min} is the lower bound of gene a_{ij}, $f(g) = r_2(1 - g/G_{max})^2$, r_2 is a random number, g is the current iteration number, G_{max} is the maximum iteration number and r is a random of [0,1].

Algorithm 2: Pseudo Code of the genetic algorithm

Input: $x_s^{(0)} = (x^{(0)}(1), x^{(0)}(2), \dots, x^{(0)}(n))$—a sequence of training data

$\hat{x}_s^{(0)} = (\hat{x}^{(0)}(l+1), \hat{x}^{(0)}(l+2), \dots, \hat{x}^{(0)}(l+n))$—a sequence of verifying data

Output: fitness_value x_b—the value with the best fitness value in the population of populations

Parameters:

Gen_{max}—the maximum number of iterations; *n*—the number of individuals

F_i—the fitness function of the individual *i*; x_i—the population *i*

g—the current iteration number of GA; d—the number of dimension

1: /***Initialize** the population of *n* individuals which are $x_i \backslash (i = 1, 2, \dots, n)$ randomly.*/

2: /***Initialize** the parameters of **GA**: Initial probabilities of crossover p_c and mutation p_m.*/

3: **FOR EACH** (*i*: $1 \le i \le n$) **DO**

4: **Evaluate** the corresponding fitness function F_i $fitness\_popu(best(idx, 1), 1)$

5: **END FOR**

6: **WHILE** ($g < Gen_{max}$) **DO FOR EACH** (*l* = 1:*n*) **DO**

7: **IF** (p_c > *rand*) **THEN**

8: /***Conduct** the crossover operation*/ $a_{kj} = a_{kj}(1-b) + a_{lj}b$ and $a_{lj} = a_{lj}(1-b) + a_{kj}b$

9: **END IF**

10: **IF** (p_m > *rand*) **THEN**

11: /***Conduct the** Mutate operation*/ $a_{ij} = \begin{cases} a_{ij} + (a_{ij} - a_{max}) * f(g), r > 0.5 \\ a_{ij} + (a_{min} - a_{ij}) * f(g), r \le 0.5 \end{cases}$

12: **END IF END FOR**

13: **FOR EACH** (*i*: $1 \le i \le n$) **DO**

14: **Evaluate** the corresponding fitness function F_i $fitness\_popu(best(idx, 1), 1)$

15: **END FOR**

16: /***Update** the best nest x_p of the *d* generation in the genetic algorithm.*/

17: **FOR EACH** (*i*: $1 \le i \le n$) **DO IF** ($F_p < F_b$) **THEN**

18: /* The global best solution can be obtained to replace the local optimal $x_b \leftarrow x_p$*/

19: **END IF END FOR END WHILE**

20: **RETURN** x_b/* The optimal solution in the global space has been obtained.*/

3.2. Data Decomposition Hybrid (DDH) Model

The time series always changes as time goes by, and such change has the features of continuity, periodicity, trend, and a certain randomness. In the previous research, no matter which models, including single model, combined model or hybrid model, they are all applied in forecasting the whole time series. Unlike the previous research, this paper proposes a data decomposition hybrid model (DDH) based on the periodicity, trend and randomness in the time series. The basic idea of DDH is to decompose the times series based on the main influencing factors. On the basis of decomposition and recombination of traditional additive model, the layer-upon-layer decreasing is applied to improve the forecasting accuracy. Then suitable models are selected to conduct the forecasting according to the data characteristics and features. The effective decomposition of data and proper forecasting models for each part can enhance the fitting performance of the model and decrease the forecasting errors to a great degree compared with conventional single forecasting methods. The detailed steps of DDH are described below (as shown in Figure 1III):

Step 1. Observe whether the time series Y_t contains trend, periodicity and randomness, and judge the applicability of the additive model and multiplicity model. In general, compared to the additive model, the multiplicity model is more suitable for time series with large fluctuations [42]. The electrical load time series have a relatively stable fluctuation range; therefore, the additive model is chosen, and the following discussion is based on it.

Step 2. Apply the moving average method or other methods to extract the periodicity C_t.

Step 3. Without the periodicity C_t, the rest of the data can be defined as trend T_t. If T_t is far larger than C_t, a periodic adjustment of C_t should be conducted to obtain the estimated periodicity $\hat{C}_t$, and this is because if we firstly forecast larger data, there will be much noise in the latter data, which will affect the forecasting accuracy. Then the new trend T_t' can be obtained ($T'_t = Y_t - \hat{C}_t$). Finally, EMD-GA-GRNN can be utilized to forecast T_t', and the forecasting value is $\hat{T}_t$. On the contrary, if C_t is far larger than T_t, EMD-GA-GRNN is used to forecast the trend T_t, and get the forecasting value $\hat{T}_t$. Then the periodicity data C_t can be obtained. Finally, the estimated value $\hat{C}_t$ is obtained through the periodic adjustment.

Step 4. The original randomness R_t is calculated ($R_t = Y_t - \hat{C}_t - \hat{T}_t$). We forecast the randomness after decomposition by applying GA-GRNN to get the forecasting value $\hat{R}_t$. The randomness after decomposition is nearly stable, so EMD is unnecessary.

Step 5. Utilize the additive model to get the final forecasting values of the time series: $\hat{Y}_t = \hat{C}_t + \hat{T}_t + \hat{R}_t$.

3.3. The EMD-GA-GRNN Forecasting Model

In the model of DDH, EMD-GA-GRNN is proposed, which is based on the data state after applying the layer-upon-layer decreasing method. However, data after the layer-upon-layer decreasing method may include some noise due to the forecasting accuracy in the former forecasting methods. Thus, it is pivotal to apply a proper method to remove the noise in the decomposed data. This paper chooses the empirical mode decomposition method considering its advantages in dealing with non-linear time series data. Then the GRNN is utilized to forecast the dealt data, because it performs well in fitting non-stationary data. The training process of GRNN is actually to ensure the optimum s, and the specific steps of the hybrid model EMD-GA-GRNN are listed as follows (Pseudo code of Algorithm 3):

Algorithm 3: Pseudo code of the hybrid model of EMD-GA-GRNN

Input: $x_s^{(0)} = \left(x^{(0)}(1), x^{(0)}(2), \ldots, x^{(0)}(q) \right)$—a sequence of training data

$\qquad x_p^{(0)} = \left(x^{(0)}(q+1), x^{(0)}(q+2), \ldots, x^{(0)}(q+d) \right)$—a sequence of verifying data

Output: $\hat{y}_z^{(0)} = \left(\hat{y}^{(0)}(q+1), \hat{y}^{(0)}(q+2), \ldots, \hat{y}^{(0)}(q+d) \right)$—forecasting electrical load from GRNN

Fitness function: $fitness = 1 / \sum\limits_{i=1}^{N} \sum\limits_{j=1}^{K} \left(Y_j(i) - \overline{Y}_j(i) \right)^2$ /*The objective fitness function*/

Parameters:

$\qquad Gen_{max}$—the maximum number of iterations; n—the number of individuals

$\qquad F_i$—the fitness function of individual i; x_i—the total population i

$\qquad G$—the current iteration number; d—the number of dimension

1: /* **Process** original electrical load time series data with the noise reduction method EMD */

2: /***Initialize** the population of n individuals x_i ($i = 1, 2, \ldots, n$) randomly.*/

3: /*Initialize the original parameters: Initial probabilities of crossover p_c and mutation p_m.*/

Algorithm 3: *Cont.*

4: FOR EACH (i: $1 \leq i \leq n$) **DO**

5: Evaluate the corresponding fitness function F_i $fitness = 1/\sum\limits_{i=1}^{N}\sum\limits_{j=1}^{K}(Y_j(i) - \overline{Y}_j(i))^2$

6: END FOR

7: WHILE ($g < Gen_{max}$) **DO**

8: FOR EACH ($i = 1{:}n$) **DO IF** ($p_c > rand$) **THEN**

9: Conduct the crossover operation of **GA** to optimize the smoothing factor of **GRNN**

10: END IF

11: IF ($p_m > rand$) **THEN**

12: Conduct the mutate operation of **GA** to optimize the smoothing factor of **GRNN**

13: END IF END FOR

14: FOR EACH (i: $1 \leq i \leq n$) **DO**

15: Evaluate the corresponding fitness function F_i $fitness = 1/\sum\limits_{i=1}^{N}\sum\limits_{j=1}^{K}(Y_j(i) - \overline{Y}_j(i))^2$

16: END FOR

17: /*Update best nest x_p of the d generation to replace the former local optimal solution.*/

18: FOR EACH (i: $1 \leq i \leq n$) **DO IF** ($F_p < F_b$) **THEN** $x_b \leftarrow x_p$;

19: END IF END FOR END WHILE

20: RETURN x_b/* **Set** the weight and threshold of the **GRNN** according to x_b.*/

21: Use x_t to train the **GRNN** and update the weight and threshold of the **GRNN** and **input** the historical data into **GRNN** to obtain the forecasting value $\hat{y}$.

Step 1. Data addressed by layer-upon-layer decreasing method would include some noises, affecting the forecasting accuracy; therefore, the first step is to denoise the composed data by using EMD method.

Step 2. Standardize and code the time series after the denoising.

Step 3. Generate the initial population $P(t)$, and the evolutionary generation is $t = 0$.

Step 4. Code the chromosome, and get the parameters of GRNN, which can be used to train the network structure.

Step 5. Set the individual evaluation standard according to the fitness function in Equation (34):

$$fitness = \frac{1}{\sum\limits_{i=1}^{N}\sum\limits_{j=1}^{K}(Y_j(i) - \overline{Y}_j(i))^2} \tag{34}$$

where $Y_j(i)$ is the output of GRNN and $\overline{Y}_j(i)$ is the output.

Step 6. Apply the optimum strategy based on the values of fitness function.

Step 7. Judge whether the fitness value meets the accuracy requirement. If so, the process ends; or move to the next step.

Step 8. Judge whether the current iteration t gets to the maximum iteration. If so, the process ends; or go to the next step.

Step 9. Perform the selection, crossover and mutation operation for the current population.

Step 10. Generate the new generation of the population, and the iteration t becomes $t + 1$, return Step 3.

4. Experiments

With the rapid development of technology and science, the electrical power system in each country tends to develop fast as well. Similarly, the power grid management has become more complicated. The forecasting is the premise and basis of decision and control; therefore, the premise and the most vital step of electrical load management is to conduct the electrical load forecasting.

The accurate forecasting can not only help the electrical power system operate safely based on reasonable maintenance schedules, but it can also decrease the grid costs and maximize the profits.

4.1. Model Evaluation

To conduct the model evaluation can lead to a clear and direct understanding of the forecasting accuracy, and it is helpful to analyze the reasons causing errors to enhance the forecasting performance. The main reasons are listed below:

(1) *Selection of influencing factors when constructing mathematical models.* In truth, the time series is affected by various factors, and it is difficult to master all of them. Therefore, errors between forecast values and actual values cannot be avoided.
(2) *Improper algorithms.* For forecasting, we just build a relatively appropriate model, so if the algorithms are chosen wrongly, the errors would become larger.
(3) *Inaccurate or incomplete data.* The forecasting should be based on the historical data, so inaccurate or incomplete data can result in forecasting errors.

When there are abnormal values, we are supposed to find the reasons causing the errors and correct each step of the model. The forecasting accuracy plays a crucial role in assessing a forecasting algorithm, and two types of evaluation metrics are chosen to evaluate the forecasting accuracy: the accuracy of forecasting a single point and the overall accuracy of forecasting multiple points. Two evaluation metrics are applied to examine a single point forecasting accuracy, which are absolute error (AE) and relative error (RE). Then we select four evaluation metrics, including mean absolute error (MAE), root mean square error (RMSE), mean absolute percentage error (MAPE) and mean error (ME), to evaluate the model performance more comprehensively. MAPE is a generally accepted metric for forecasting accuracy, and MAE and RMSE can measure the average magnitude of the forecast errors; however, RMSE imposes a greater penalty on a large error than several small errors [43].

For a group of time series x_t ($t = 1, 2, \ldots, T$), the corresponding forecasting output is $\hat{x}_t$ and detailed description of evaluation metrics is shown in Table 1.

Table 1. The evaluation metrics.

Name of Metrics	Equation	No.	Name of Metrics	Equation	No.
MAE	$MAE = \frac{1}{T} \sum_{t=1}^{T} |x_t - \hat{x}_t|$	(35)	ME	$ME = \frac{1}{T} \sum_{t=1}^{T} (x_t - \hat{x}_t)$	(38)
RMSE	$RMSE = \sqrt{\frac{1}{T} \sum_{t=1}^{T} (x_t - \hat{x}_t)^2}$	(36)	AE	$AE = x_t - \hat{x}_t$	(39)
MAPE	$MAPE = \frac{1}{T} \sum_{t=1}^{T} \left| \frac{x_t - \hat{x}_t}{x_t} \right| \times 100\%$	(37)	RE	$RE = \frac{x_t - \hat{x}_t}{x_t}$	(40)

The smaller values of the six metrics are, the higher forecasting accuracy is. Therefore, the evaluation metrics can both reflect the forecasting results and its accuracy clearly and directly and provide a reference base for decisions, which is beneficial to improving the model and conducting the analysis. Thus, the significance of the evaluation metrics is very large.

4.2. Experimental Setup

This paper uses the 30-min interval data of New South Wales, Australia in April 2011 to verify the effectiveness of the proposed hybrid DDH model based on data decomposition. In the first experiment, the data size is 1440, and data in the first 29 days are the training set, and the testing set includes data in the 30th day. The detailed ideas of the proposed electrical load hybrid model is summarized as follows (as shown in Figure 2):

(1) The original electrical load time series data Y_t has an obvious trend and periodicity. Initially, the moving average method is conducted to extract the periodicity C_t. For the periodicity C_t, conduct the periodic adjustment and obtain $\hat{C}_t$.

(2) Subtract the periodicity of the original time series data, and get the original trend $T_t(T_t = Y_t - \hat{C}_t)$. For the original data without periodicity, EMD needs to be initially applied to eliminate the noises and improve the forecasting accuracy. Then the genetic algorithm could be used to optimize GRNN to obtain the forecasting trend item $\hat{T}_t$.

(3) Finally, the randomness can be obtained through $R_t = Y_t - \hat{C}_t - \hat{T}_t$, then the GRNN optimized by the genetic algorithms is utilized to forecast the randomness and the forecasting value is obtained. The trend tends to be steady; therefore, there is no need to eliminate noises.

(4) The final forecasting is performed by the additive model of time series $\hat{Y}_t = \hat{C}_t + \hat{T}_t + \hat{R}_t$.

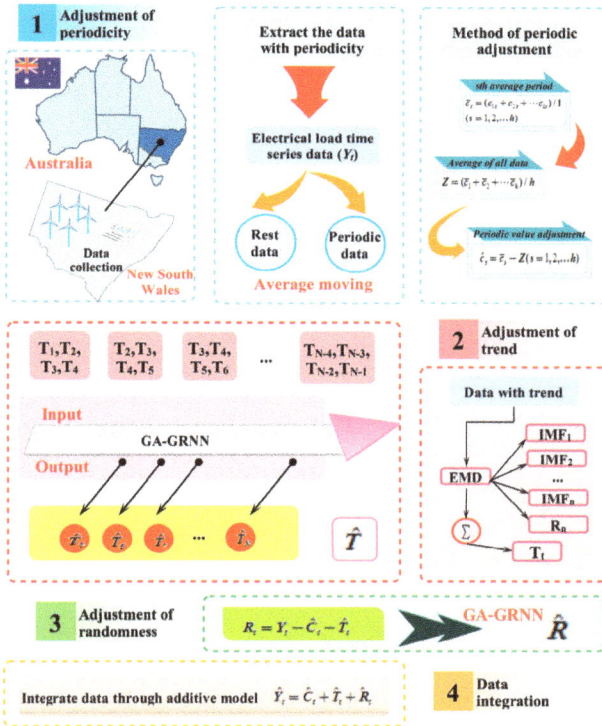

Figure 2. The process of electrical load forecasting for New South Wales.

4.3. Empirical Results

The model performance is evaluated based on the upper data, and the results are obtained by using MATLAB®(2015a), which was implemented under Windows 8.1 with a 2.5 GHz Intel Core i5 3210 M, 64 bit CPU with 4 GB RAM. Figure 3 shows the data decomposition process.

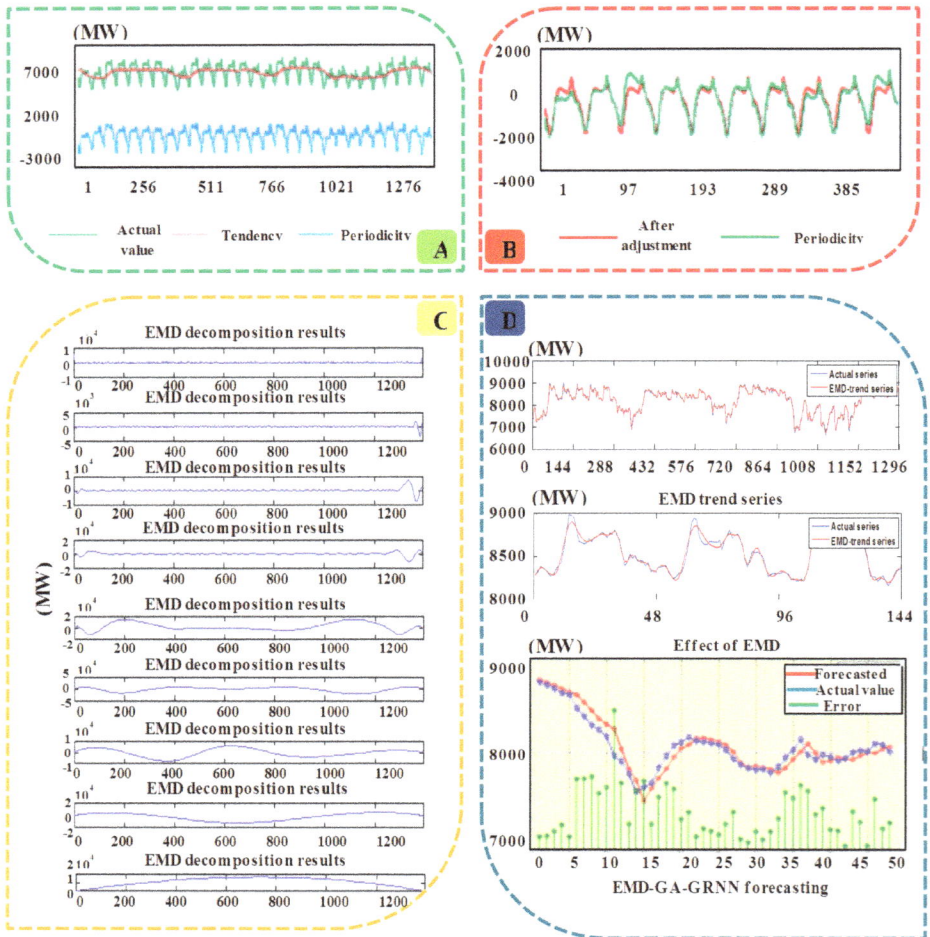

Figure 3. The forecasting effects. (**A**) The original electrical load time series; (**B**) Electrical load time series data after adjustment; (**C**) EMD decomposition results; (**D**) EMD trend series, effect of EMD and results of EMD-GA-GRNN forecasting.

(1)   Figure 3A shows the results after decomposition by moving average, from which it can be seen that the original electrical load data contains a certain periodicity, and the variation of the period is roughly equal, so the additive model is more suitable. The length of the period $h = 48$ can be ensured based on the data distribution. Thus the moving average method is used to decompose the electrical load data into two parts, which are periodicity and trend. Besides, from the decomposed results, it can be known that the level of trend is nearly ten times the periodicity. This is because that the moving average method can demonstrate the large trend of the development, eliminating the fluctuation factors such as season. Therefore, the periodic adjustment should be conducted through extracting the periodicity.

(2)   Figure 3B is the electrical load data after periodic adjustment, from which is can be known that the electrical load data after the periodic adjustment have periodic sequence and basis trend characteristics.

(3) Figure 3C demonstrates the output results of trend data after EMD. It shows that nine components are obtained, including IMF_1, IMF_2, ..., IMF_8 and R_n, after EMD data decomposition. The high-frequency data in highest component is removed, and the rest data are regarded as the new trend time series data.

(4) Figure 3D clearly reveals the trend data after EMD decomposition by removing the high frequency component, and it can be obviously seen that the data denoised by EMD are smoother than the original data.

Next, data after removing the high frequency component by EMD is fitted and forecast by GRNN. The genetic algorithm is applied to optimize the smoothing factor σ in GRNN. The hybrid electrical load forecasting model EMD-GA-GRNN constructed in this paper is applied to forecast the trend value in the next time point by using the historical data in the past time point. In this experiment, the trend value of the former four time points are used to forecast the trend value of the 5th time point. For the given data, the data need initially to be divided into the training sample and testing sample. Take the training sample for example, x_1, x_2, x_3, x_4, x_5 is the first sample group, and x_1, x_2, x_3, x_4 are independent variables, and x_5 is the objective function value. Similarly, x_2, x_3, x_4, x_5, x_6 is the second sample group, x_2, x_3, x_4, x_5 are independent variables, and x_6 is the objective function value. By that analogy, the final training matrix is:

$$\begin{pmatrix} x_1 & x_2 & x_3 & \cdots & x_{1292} \\ x_2 & x_3 & x_4 & \cdots & x_{1293} \\ x_3 & x_4 & x_5 & \cdots & x_{1294} \\ x_4 & x_5 & x_6 & \cdots & x_{1295} \\ x_5 & x_6 & x_7 & \cdots & x_{1296} \end{pmatrix} \tag{41}$$

where each column is a sub-sample sequence, and the last row is the expected output. The training sample is used to train GA-GRNN, after that the network after training is obtained. The forecasting effects can be clearly seen from Figure 3D that EMD-GA-GRNN has a better fitting effect, and MAPE between network output and real value is 2.11%. The training model in Figure 4 is shown as follows.

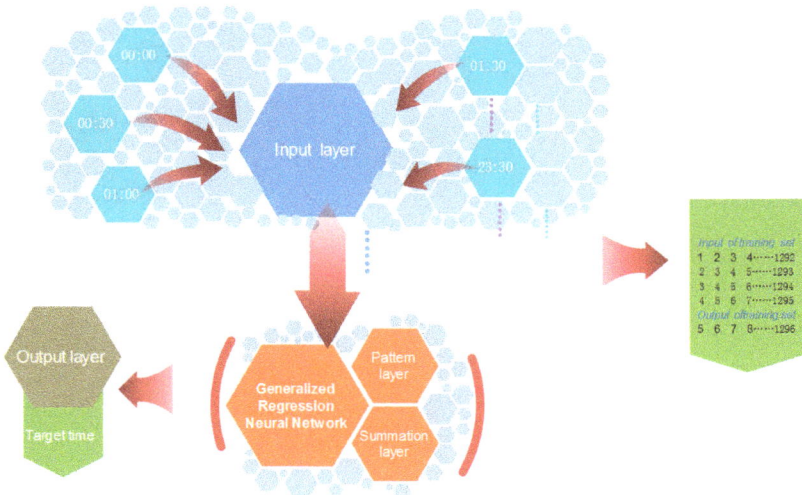

Figure 4. The generalized regression neural network model.

To the best of our knowledge, a great variety of forecasting approaches can achieve good performance in dealing with non-linear time series; therefore, in this paper we compared the proposed GRNN with three other well-known and commonly used methods, including wavelet neural network (WNN), the secondary exponential smoothing method (SES) and auto regressive integrated moving average (ARIMA). The forecasting results are compared as shown in Figure 5, from which it can be known that:

(1) The speed to forecast the nonlinear time series data by using WNN is fast, with a better ability of generalization and a higher accuracy; however, the stability is weak.

(2) The advantages of SES are the simple calculation, strong adaptability and stable forecasting results, but the ability to address nonlinear time series data is weak.

(3) ARIMA performs well with a relatively higher accuracy when forecasting the electrical load data. However, as time goes by, the forecasting errors would gradually become larger and larger, which is only suitable for short-term forecasting.

(4) On the whole, compared with other methods, GRNN can obtain a better and more stable forecasting result, as it deals with the non-linear data well and can fit and forecast the electrical load data well.

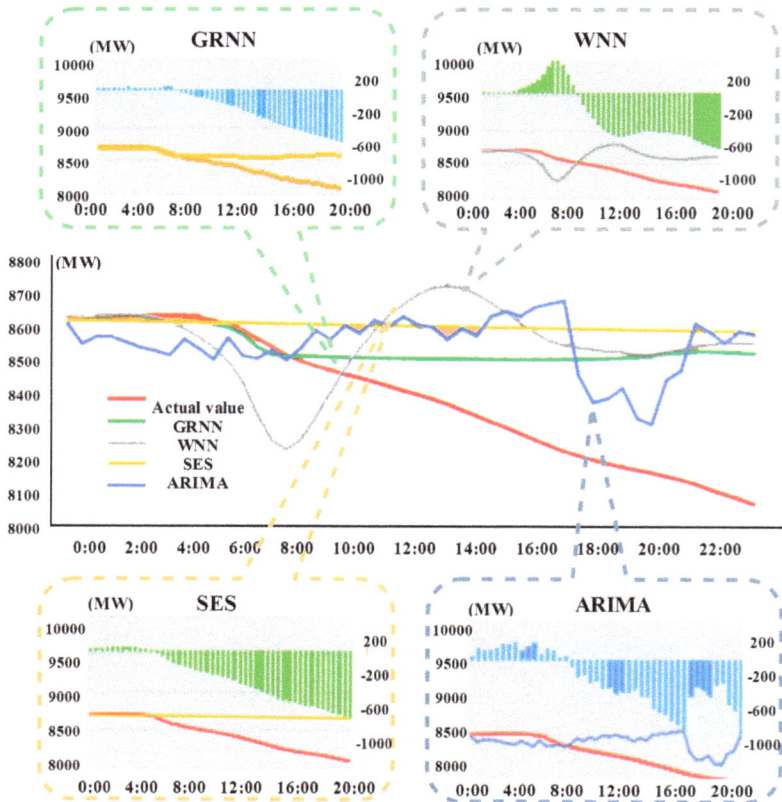

Figure 5. Forecasting results for trend of each model after removing the periodicity.

Next the randomness is obtained by $R_t = Y_t - \hat{C}_t - \hat{T}_t$. Because it tends to be stationary, we can only apply GA-GRNN to get the forecasting value $\hat{R}_t$. The forecasting results of DDH can be calculated by the additive model $\hat{Y}_t = \hat{C}_t + \hat{T}_t + \hat{R}_t$, and results are shown in Figure 6. Figure 6II demonstrates that the forecasting error in the 11th time point is the largest with an MAPE within 5%, and this results is satisfactory.

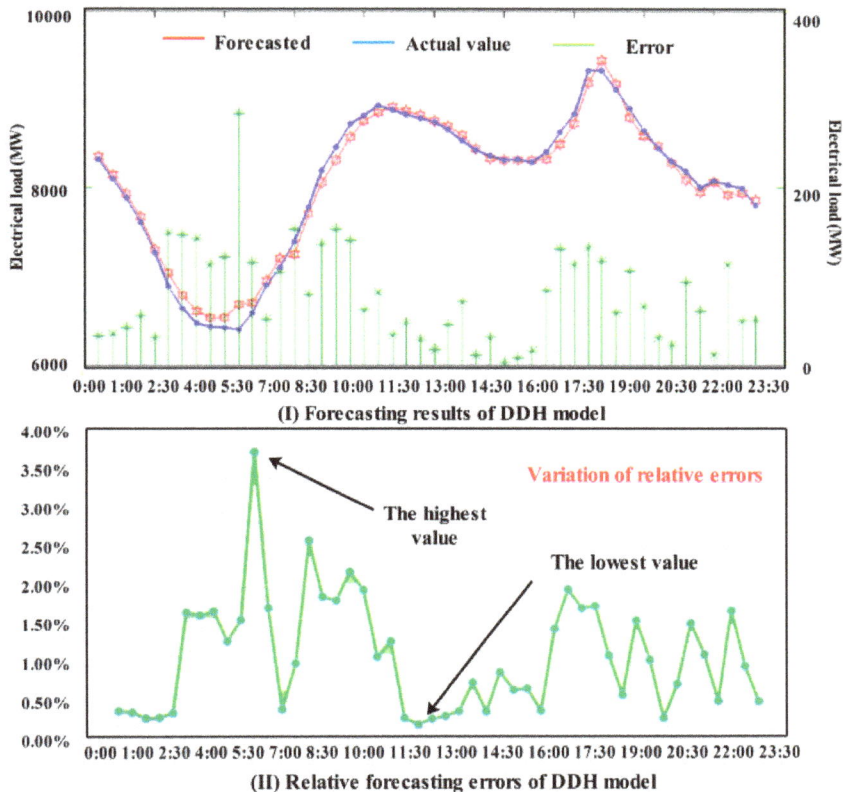

Figure 6. Forecasting results and MAPE of DDH model.

4.4. Comparative Analysis

In order to prove the good performance of the proposed DDH model in this paper, three other hybrid models are compared with it, which are EMD-GA-WNN, GA-GRNN and EMD-GA-GRNN. The comparison results are shown in Table 2.

(1) From Figure 7, it can be seen that EMD-GA-WNN does not perform well when forecasting the electrical load data, and the relative errors of some parts even exceed 5%. This may be caused by the weak forecasting stability of WNN, and although GA can optimize its parameters, the effect to improve its stability is weak.

(2) As for GA-GRNN and EMD-GA-GRNN, MAPEs are all within 5%, which indicates that the two forecasting models have better performance. In detail, the forecasting effect of EMD-GA-GRNN is much better than that of GA-GRNN, proving the function of EMD in improving the forecasting accuracy.

(3)   The DDH model based on the data decomposition put forward in this paper can control the MAPE at 4%; thus, it can be known that it has a very strong fitting ability for non-linear data and forecasting ability for the electrical load time series. Both the simulation results and the forecasting process demonstrate that the proposed model can have a good performance when forecasting the non-linear time series data with periodicity, trend and randomness.

(4)   From the evaluation metrics in Figure 7, it can be known that the forecasting ability of GRNN is better than WNN, which is because that GRNN can deal well with the data such as electrical load time series; therefore, this paper also establishes the model based on GRNN. The proposed forecasting model EMD-GA-GRNN and EMD-GA-GRNN based on WNN and GRNN can improve the forecasting accuracy well. However, in comparison, GRNN is more suitable for the nonlinear time series data, and MAPEs of EMD-GA-WNN and EMD-GA-GRNN are 2.22% and 1.53%, respectively. Certainly, EMD can reduce the forecasting errors in some degree. Besides, MAPE decreases from 1.62% of GA-GRNN to 1.53% of EMD-GA-GRNN. However, DDH model can reduce MAPE within 1%.

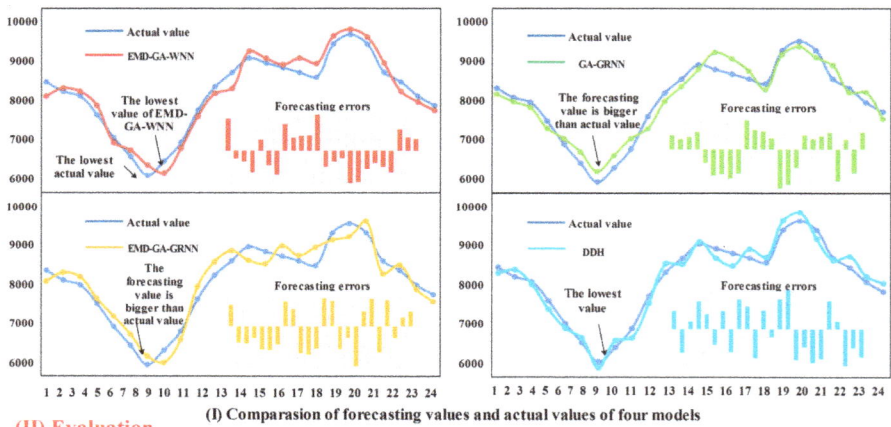

(I) Comparasion of forecasting values and actual values of four models

(II) Evaluation metrics

$$MAE = \frac{1}{T}\sum_{t=1}^{T}|x_t - \hat{x}_t|$$

$$RMSE = \sqrt{\frac{1}{T}\sum_{t=1}^{T}(x_t - \hat{x}_t)^2} \quad ME = \frac{1}{T}\sum_{t=1}^{T}|x_t - \hat{x}_t|$$

$$MAPE = \frac{1}{T}\sum_{t=1}^{T}\left|\frac{x_t - \hat{x}_t}{x_t}\right| \times 100\% \quad RE = \frac{(x_t - \hat{x}_t)}{x_t}$$

$$AE = x_t - \hat{x}_t$$

(III) MAE, RMSE, MAPE and ME of four models

	MAE	RMSE	MAPE	ME
EMD-GA-WNN	168.1870	217.1040	2.22%	-147.5324
GA-GRNN	130.9404	167.7759	1.62%	-59.7562
EMD-GA-GRNN	124.0530	163.6047	1.53%	-26.5351
DDH	77.0542	97.7689	0.98%	28.6760

Figure 7. Forecasting results of each model.

Table 2. The forecasting output of each model.

| Time | Actual Value | Forecasting Output of the Model | | | | Time | Actual Value | Forecasting Output of the Model | | | |
		EMD-GA-WNN	GA-GRNN	EMD-GA-GRNN	DDH			EMD-GA-WNN	GA-GRNN	EMD-GA-GRNN	DDH
0:00	8314.34	8316.79	8485.70	8448.82	8297.61	12:00	8731.01	8911.29	8740.80	8749.89	8719.90
0:30	8097.56	8120.24	8324.54	8277.18	8082.89	12:30	8649.94	8869.53	8629.02	8653.62	8667.12
1:00	7881.03	7928.45	8120.26	8077.17	7873.46	13:00	8520.88	8820.83	8534.06	8544.21	8572.19
1:30	7613.53	7687.64	7880.49	7856.92	7621.62	13:30	8419.95	8746.33	8436.37	8418.62	8403.67
2:00	7265.94	7330.03	7586.18	7558.97	7253.44	14:00	8359.06	8668.88	8332.43	8324.83	8296.61
2:30	6884.16	7067.63	7116.65	7134.47	6993.73	14:30	8304.55	8570.31	8261.18	8263.58	8263.13
3:00	6638.55	6870.98	6725.93	6750.43	6741.38	15:00	8312.49	8529.19	8219.80	8205.82	8269.39
3:30	6464.02	6743.91	6519.60	6508.65	6567.98	15:30	8285.25	8513.52	8219.06	8196.43	8268.39
4:00	6421.67	6707.56	6392.63	6404.61	6497.16	16:00	8401.56	8545.67	8210.85	8185.83	8287.56
4:30	6412.23	6724.73	6419.99	6451.49	6507.20	16:30	8616.59	8660.40	8297.24	8257.93	8451.27
5:00	6394.72	6873.82	6536.38	6556.23	6641.75	17:00	8829.22	8806.80	8600.25	8503.72	8683.46
5:30	6582.09	7033.42	6598.89	6605.74	6691.06	17:30	9308.85	9216.90	9000.54	8890.66	9152.10
6:00	6899.25	7399.88	6802.04	6814.45	6914.60	18:00	9307.55	9429.90	9249.77	9200.67	9398.89
6:30	7093.06	7642.45	7200.95	7193.05	7155.01	18:30	9106.91	9271.13	9328.58	9330.35	9145.80
7:00	7395.69	7715.40	7589.45	7524.90	7203.21	19:00	8893.08	8987.73	9292.04	9240.84	8762.06
7:30	7783.28	7973.56	7850.94	7754.65	7642.06	19:30	8641.32	8747.35	9042.83	8970.91	8562.08
8:00	8193.10	8145.44	8135.61	8026.69	8048.81	20:00	8437.27	8565.72	8596.66	8541.77	8445.35
8:30	8454.25	8325.99	8475.43	8493.17	8270.93	20:30	8297.38	8389.37	8407.47	8394.00	8249.05
9:00	8710.86	8616.62	8718.43	8606.26	8544.31	21:00	8174.13	8189.12	8278.29	8260.20	8057.12
9:30	8806.93	8824.25	8975.41	8860.02	8722.24	21:30	8004.14	7998.97	8129.02	8116.55	7925.10
10:00	8920.64	8984.50	8994.28	8924.99	8815.63	22:00	8077.99	8098.08	7964.69	7949.84	8050.27
10:30	8872.57	9023.17	9034.39	8910.56	8863.76	22:30	8033.84	7965.71	7977.29	7939.60	7904.37
11:00	8816.19	8981.31	8953.15	8840.24	8815.61	23:00	7989.55	7951.66	8016.42	7970.00	7923.18
11:30	8777.05	8946.75	8844.91	8810.79	8769.22	23:30	7803.73	7841.56	8020.16	7972.34	7829.81

The summary is concluded in Remark 1.

Remark 1. *It can be concluded that compared to the single forecasting model, DDH model is more suitable for forecasting the electrical load time series data with a higher fitting ability and better forecasting capacity.*

The analysis above only shows results of three models in one experiment, but it cannot comprehensively and fully demonstrate the model performance. Each model will be trained 10 times with the same iteration numbers to make the forecasting results more stable. The obtained forecasting quality and results are shown in Figure 8 and Table 3. The two figures both indicate that DDH model based on the data decomposition perform well when measured by different evaluation metrics. A smaller MAE means a higher forecasting accuracy, a lower RMSE indicates a better fitting degree of electrical load, and MAPE is an index to assess the forecasting ability of the model. At present, for the data of New South Wales, the best standard is about 1%. From the average of MAE in ten experiments, DDH has the smallest value, indicating the best forecasting accuracy. What is more, the smallest RMSE cannot only mean that DDH can fit the electrical load time series well, but it can also prove that the forecasting results of the model are stable.

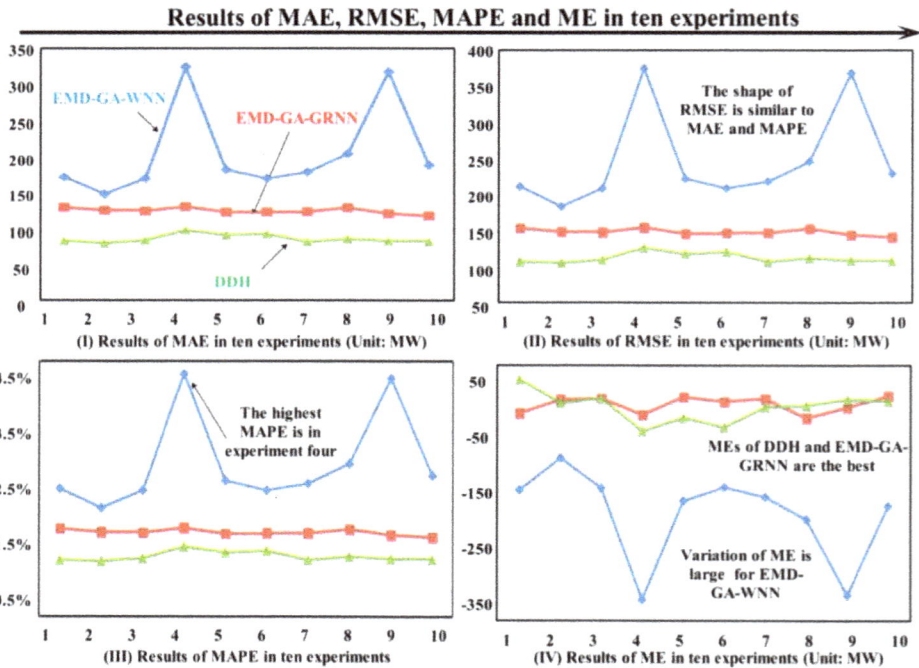

Figure 8. Model evaluation of three forecasting models.

Furthermore, MAPE of DDH also shows that DDH model based on the data decomposition put forward in this paper can reach the best forecasting standard currently.

Table 3. Forecasting performance evaluation results.

No.	EMD-GA-WNN				EMD-GA-GRNN				DDH			
	MAE	RMSE	MAPE	ME	MAE	RMSE	MAPE	ME	MAE	RMSE	MAPE	ME
1	168.9702	218.9832	0.0223	−150.3171	124.0530	163.6047	0.0153	−26.5351	77.0542	97.7687	0.0098	28.6760
2	143.4750	187.9665	0.0189	−97.7305	120.0837	160.1870	0.0147	−4.4056	73.8617	90.7383	0.0096	−9.6550
3	166.6531	215.6456	0.0220	−146.7813	119.0237	161.9871	0.0146	−2.3599	77.7576	96.1427	0.0101	−2.7978
4	327.7693	366.7733	0.0423	−327.7693	125.1301	160.2820	0.0154	−28.9499	91.7547	110.8149	0.0121	−54.7912
5	179.9683	231.0331	0.0237	−167.4704	117.3492	158.6897	0.0144	−0.0239	85.2639	101.5193	0.0111	−33.4767
6	166.7894	215.3270	0.0220	−145.6936	117.6313	157.8023	0.0145	−7.9405	86.7251	105.4555	0.0114	−48.7363
7	176.4214	228.1589	0.0232	−161.6078	117.9699	158.3187	0.0145	−3.1824	75.6047	90.9116	0.0098	−16.3685
8	203.5435	250.7716	0.0267	−197.7634	123.8518	163.8211	0.0152	−34.5862	79.7625	96.8084	0.0104	−14.4799
9	320.8268	358.9548	0.0415	−320.8268	115.4209	157.6318	0.0142	−16.9704	77.0379	94.9811	0.0100	−4.2135
10	186.7493	239.4212	0.0246	−176.1809	112.3399	155.0562	0.0138	1.6029	75.9955	93.4155	0.0099	−6.9212
Mean	204.1166	251.3035	0.0267	−189.2141	119.2854	159.7381	0.0147	−12.3351	80.0818	97.8556	0.0104	−16.2764

4.5. Further Experiments

Initially, in order to further prove the effectiveness of the proposed DDH hybrid model, we expand our sample size by using the data in 89 days to forecast the data in the 90th day. That is to say, the first 89th days are the training set, and the testing set include data in the 90th day. The experiment results of both working days and weekends are shown in Table 4. Besides, experiments of days in different seasons are also done to examine the effectiveness and robustness of the proposed hybrid model, which are listed in Table 4 and detailed analysis are as follows:

(1)   As for the weekly analysis, it can be seen that the average MAPE of DDH in one week is 1.01%, which is lower than EMD-GA-WNN and EMD-GA-GRNN. About other indexes, including MAE, RMSE and ME, DDH all obtain the best forecasting results. When comparing the working days with weekends, the proposed hybrid model can both have a high forecasting accuracy, which proves the effectiveness of the model.

(2)   Table 5 shows the forecasting results of days in different seasons. Based on the comparison, it can be concluded that DDH is superior to the other two models with the values of MAPE 0.96%, 1.18%, 1.18% and 1.13% in spring, summer, autumn and winter, respectively. The results can validate that the proposed hybrid DDH model has a high degree of robustness and forecasting accuracy.

The summary is concluded in Remark 2.

Remark 2. *The performance of the DDH model is stable and good when forecasting the electrical load data in one week and different seasons.*

Table 4. Forecasting performance evaluation results of one week with larger training set.

	Monday	Tuesday	Wednesday	Thursday	Friday	Saturday	Sunday	Week
				EMD-GA-WNN				
MAE	117.6839	129.4867	142.1135	279.4283	225.6017	118.0649	206.1598	174.0770
RMSE	288.4620	237.0900	218.0418	211.3369	202.1987	245.0624	289.4222	241.6591
MAPE	0.0249	0.0286	0.0273	0.0263	0.0275	0.0219	0.0226	0.0256
ME	−129.1364	−18.7732	−130.2645	−88.3712	−32.1989	−60.3024	−115.4213	−82.0668
				EMD-GA-GRNN				
MAE	117.6850	126.4213	110.0889	112.1325	129.0178	157.3144	138.0976	127.2511
RMSE	141.7699	168.3712	155.2626	148.0987	161.5546	132.1019	168.3174	153.6395
MAPE	0.0144	0.0137	0.0149	0.0155	0.0158	0.0159	0.0142	0.0149
ME	−98.6273	−87.0125	−110.2455	−136.4188	−65.231	−21.0987	−33.4685	−78.8718
				DDH				
MAE	76.4219	88.1348	76.1653	79.0187	84.315	69.1083	70.4245	77.6555
RMSE	97.6681	102.4269	99.8349	105.1917	112.3416	108.1947	98.1032	103.3944
MAPE	0.0101	0.0094	0.0112	0.0095	0.0098	0.0106	0.0103	0.0101
ME	−13.0719	−2.0715	−4.3728	18.1605	12.1004	−34.5671	−10.0628	−4.8407

Table 5. Forecasting performance evaluation results of different seasons with a larger training set.

Evaluation Index	EMD-GA-WNN	EMD-GA-GRNN	DDH	Evaluation Index	EMD-GA-WNN	EMD-GA-GRNN	DDH
	Spring				Summer		
MAE	119.4287	117.0216	76.0138	MAE	137.0345	108.417	60.1837
RMSE	292.0655	140.3726	97.0138	RMSE	213.0418	156.1783	94.1296
MAPE	0.0231	0.0158	0.0096	MAPE	0.0274	0.0162	0.0118
ME	−112.0659	−78.4257	−16.1076	ME	−97.3125	−52.1035	−20.0244
	Autumn				Winter		
MAE	125.0638	100.0246	78.1025	MAE	112.0605	100.4629	73.1068
RMSE	213.1294	148.7329	89.4237	RMSE	200.0217	158.0376	971136
MAPE	0.0219	0.0143	0.0118	MAPE	0.0212	0.0158	0.0113
ME	−101.0137	−25.4269	−11.0036	ME	−94.1346	−36.0599	−11.0217

In addition, we also compare the forecasting performance of the proposed DDH model in this paper to the models in the literature, including [1,4,44,45]. As shown in Table 6, the model in this paper improves the forecasting accuracy by 0.089% compared to the HS-ARTMAP network. The MAPEs of the combined model based on BPNN, ANFIS and diff-SARIMA and hybrid model based on WT, ANN and ANFIS are 1.654% and 1.603%, respectively. In the compared models, the combined model based on BPNN, RBFNN, GRNN and GA-BPNN has the lowest MAPE, which is 1.236%. Therefore, in summary, the DDH model outperforms the other compared models in the literature. The superior performance of DDH is because that the model can deal with both trend and periodicity in the original time series, which can greatly enhance the forecasting accuracy. Besides, compared to conventional BPNN and ARIMA, GRNN has a strong ability of generalization, robustness, fault tolerance and convergence ability.

Table 6. Comparison of MAPE with models in the literature.

Model	Period	MAPE (%)	Ref.
Combined model based on BPNN, ANFIS and diff-SARIMA	Data from May to June 2011	1.654	[1]
Combined model based on BPNN, RBFNN,GRNN and GA-BPNN	Data from February 2006 to February 2009	1.236	[4]
HS-ARTMAP network	Data in the head days in January from 1999 to 2009	1.900	[44]
Hybrid model based on WT, ANN and ANFIS	Data from 12 July to 31 July 2004	1.603	[45]
The proposed DDH	Data from April to June 2011	1.010	/

4.6. Discussion on Model Features

As discussed above, the major model in DDH model is GRNN which is optimized by GA. The experimental results also demonstrate their effectiveness in forecasting the short-term electrical load time series. This part will discuss the advantages of GRNN and GA further and more deeply. As shown in Table 7, GRNN has four obvious features:

1. It has a relatively low requirement for the sample size during the model building process, which can reduce the computing complexity;
2. The human error is small. Compared with the back propagation neural network (BPNN), GRNN is different. During the training process, the historical samples will directly control the learning process without adjusting the connection weight of neurons. What is more, parameters like learning rate, training time and the type of transfer function, need to be adjusted. Accordingly, there is only one parameter in GRNN that needs to be set artificially, which is the smoothing factor;
3. Strong self-learning ability and perfect nonlinear mapping ability. GRNN belongs to a branch of RBF neural networks with strong nonlinear mapping function. To apply GRNN in electrical load forecasting can better reflect the nonlinear mapping relationship;
4. Fast learning rate. GRNN uses BP algorithm to modify the connection weight of the relative network, and applies the Gaussian function to realize the internal approximation function, which can help arrive at an efficient learning rate. The above features of GRNN play a pivotal role in performing the electrical load forecasting when the original data are fluctuating and non-linear.

Table 7. Comparison of GRNN with grey model, regression model and BP model.

Model	Nonlinear Mapping Ability	Set parameters Artificially	Generalization	Robustness	Fault Tolerance	Structural Interpretability	Convergence Ability	Sample Size
Grey model	Middle	Middle	Weak	Weak	Weak	Good interpretability for internal structure	—	Large
Regression model	Weak	Large	Middle	Weak	Weak	Good interpretability for internal structure	—	Large
Back Propagation (BP) model	Strong	Large	Middle	Middle	Weak	No structural interpretability	Weak	Large
GRNN model	Strong	Only smoothing factor parameter	Strong	Strong	Strong	No structural interpretability	Strong	Low requirement

The genetic algorithm is utilized to optimize the only one parameter in GRNN, and it is a type of algorithm that works without limiting the field or type of the problem. That is to say, it does not depend on detailed problems, and can provide a universal framework to solve problems. Compared to the traditional optimization, it has the following advantages:

- Self-adaptability. When solving problems, GA deals with the chromosome individuals through coding. During the process of evolution, GA will search the optimal individuals based on the fitness function. If the fitness value of chromosome is large, it indicates a stronger adaptability. It obeys the rules of survival of the fittest; meanwhile, it can keep the best state in a changing environment;
- Population search. The conventional methods usually search for single points, which is easily trapped into a local optimum if a multimodal distribution exists in the search space. However, GA can search from multiple starting points and evaluate several individuals at the same time, which makes it achieve a better global searching;
- Need for a small amount of information. GA only uses the fitness function to evaluate the individuals without referring to other information. It has a small dependence or limitation conditions to the problems, so it has a wider applicability;
- Heuristic random search. GA highlights the probability transformation instead of the certain transformation rule;
- Parallelism. On the one hand, it can search multiple individuals in the solution space; on the other hand, multiple computers can be applied to perform the evolution calculation to choose the best individuals until the computation ends. The above advantages make GA widely used in many fields, such as function optimization, production dispatching, data mining, forecasting for electrical load and so on.

5. Conclusions

The electrical load forecasting can not only provide the electricity supply plans for regions in a timely and reliable way, but it can also help maintain normal social production and life. Thus, to improve the forecasting accuracy of electrical load can lower risks, improve the economic benefits, decrease the costs of generating electricity, enhance the safety of electrical power systems and help policy makers make better action plans. Therefore, how to forecast the changing trends and features of electrical loads in the power grid accurately and effectively has become a both significant and challenging problem. This paper proposes a Data Decomposition Hybrid (DDH) model based on the data decomposition that can deal well with the task, and it mainly contains two key steps:

The first one is to decompose the data based on the main factors of electrical load time series data. On the basis of decomposition and reconstitution of traditional time series additive model, the layer-upon-layer decreasing decomposition is applied for the reconstitution to enhance the forecasting accuracy. Then according to the characteristics of the decomposed data, suitable forecasting models are found to fit and forecast the sub-sequence. Through the effective decomposition of electrical load time series data and selection of proper forecasting models, the fitting ability and forecasting capacity can be well improved.

The second idea is to improve the forecasting accuracy of Generalized regression neural network (GRNN). The major forecasting model in this paper is GRNN, and genetic algorithm is utilized to optimize parameters in GRNN. Before that EMD is applied to eliminate the noises in the data. Thus, with the help of EMD and GA, the forecasting performance of GRNN can be greatly enhanced.

The experimental results show that compared with EMD-GA-WNN, GA-GRNN and EMD-GA-GRNN, the proposed hybrid model has a good forecasting effect for electrical load time series data with periodicity, trend and randomness. In practice, the DDH model based on data decomposition can reach a high forecasting accuracy, becoming a promising method in the future. Besides, if the time series show an obvious periodicity, trend and randomness, the hybrid model can be applied commonly and effectively in other forecasting fields, such as product sales forecasting, tourism demand forecasting, warning and forecasting of flood, wind speed forecasting, traffic flow forecasting and so on.

However, with the development of technology and information, there are still many problems existing in the forecasting field. This paper mainly focuses on the study of a hybrid forecasting model based on time series decomposition and how to improve the forecasting accuracy, and further analysis can be conducted in the following aspects: (1) This paper ignores the influences of other factors on the electrical time series owing to the limitations of data collection; therefore, how to design a forecasting model and algorithm of multiple variables is a problem worth studying; (2) The forecasting techniques continue to improve, and there is no a perfect forecasting model that can deal well with all time series forecasting problems. Thus, it is necessary to develop new algorithms to achieve the future forecasting work; (3) Denoising of time series. The EMD method applied in this paper is just one type of denoising method, and other algorithms, such as Kalman filtering and wavelet packet decomposition, should be compared to EMD to select a better one.

Acknowledgments: This research was supported by the National Natural Science Foundation of China (Grant No. 71671029).

Author Contributions: Yuqi Dong proposed the concept of this research and made overall guidance; Xuejiao Ma completed the whole paper; Chenchen Ma made tables and figures; Jianzhou Wang provided related data.

Conflicts of Interest: The authors declare no conflict of interest.

Abbreviations

ANNs	Artificial neural networks
SVM	Support vector machine
EA	Evolution algorithms
ARIMA	Auto regressive integrated moving average
SOM	Self-organizing map
GRNN	Generalized regression neural network
EMD	Empirical mode decomposition
IMF	Intrinsic mode function
WNN	Wavelet neural network
SES	Secondary exponential smoothing
ANFIS	Adaptive network-based fuzzy inference system
RBFNN	Radial basis function neural network
HS-ARTMAP	Hyper-spherical ARTMAP network
RBF	Radial basis function
GA	Genetic algorithm
DDH	Data Decomposition Hybrid Model
MAE	Mean absolute error
RMSE	Root mean square error
MAPE	Mean absolute percentage error
ME	Mean error
AE	Absolute error
RE	Relative error
BPNN	Back propagation neural network
diff-SARIMA	Difference seasonal autoregressive integrated moving average
ART	Adaptive resonance theory
WT	Wavelet transform

References

1. Yang, Y.; Chen, Y.H.; Wang, Y.C.; Li, C.H.; Li, L. Modelling a combined method based on ANFIS and neural network improved by DE algorithm: A case study for short-term electricity demand forecasting. *Appl. Soft Comput.* **2015**, *49*, 663–675. [CrossRef]
2. Li, S.; Goel, L.; Wang, P. An ensemble approach for short-term load forecasting by extreme learning machine. *Appl. Energy* **2016**, *170*, 22–29. [CrossRef]
3. Takeda, H.; Tamura, Y.; Sato, S. Using the ensemble Kalman filter for electricity load forecasting and analysis. *Energy* **2016**, *104*, 184–198. [CrossRef]
4. Xiao, L.Y.; Wang, J.Z.; Hou, R.; Wu, J. A combined model based on data pre-analysis and weight coefficients optimization for electrical load forecasting. *Energy* **2015**, *82*, 524–549. [CrossRef]
5. Ren, Y.; Suganthan, P.N.; Srikanth, N.; Amaratunga, G. Random vector functional link network for short-term electricity load demand forecasting. *Inf. Sci.* **2016**, *1*, 1078–1093. [CrossRef]
6. Xiao, L.Y.; Shao, W.; Liang, T.L.; Wang, C. A combined model based on multiple seasonal patterns and modified firefly algorithm for electrical load forecasting. *Appl. Energy* **2016**, *167*, 135–153. [CrossRef]
7. Xiao, L.Y.; Shao, W.; Wang, C.; Zhang, K.Q.; Lu, H.Y. Research and application of a hybrid model based on multi-objective optimization for electrical load forecasting. *Appl. Energy* **2016**, *180*, 213–233. [CrossRef]
8. Jiang, P.; Ma, X.J. A hybrid forecasting approach applied in the electrical power system based on data preprocessing, optimization and artificial intelligence algorithms. *Appl. Math. Model.* **2016**, *40*, 10631–10649. [CrossRef]
9. Azadeh, A.; Ghaderi, S.F.; Sohrabkhani, S. A simulated-based neural network algorithm for forecasting electrical energy consumption in Iran. *Energy Policy* **2008**, *36*, 37–44. [CrossRef]
10. Chang, P.C.; Fan, C.Y.; Lin, J.J. Monthly electricity demand forecasting based on a weighted evolving fuzzy neural network approach. *Electr. Power Syst. Res.* **2011**, *33*, 17–27. [CrossRef]
11. Wang, J.Z.; Chi, D.Z.; Wu, J.; Lu, H.Y. Chaotic time serious method combined with particle swarm optimization and trend adjustment for electricity demand forecasting. *Expert Syst. Appl.* **2011**, *38*, 19–29. [CrossRef]
12. Ghanbari, A.; Kazami, S.M.; Mehmanpazir, F.; Nakhostin, M.M. A cooperative ant colony optimization-genetic algorithm approach for construction of energy demand forecasting knowledge-based expert systems. *Knowl. Based Syst.* **2013**, *39*, 194–206. [CrossRef]
13. Zhao, H.R.; Guo, S. An optimized grey model for annual power load forecasting. *Energy* **2016**, *107*, 272–286. [CrossRef]
14. Zeng, B.; Li, C. Forecasting the natural gas demand in China using a self-adapting intelligent grey model. *Energy* **2016**, *112*, 810–825. [CrossRef]
15. Kavaklioglu, K. Modeling and prediction of Turkeys electricity consumption using Support Vector Regression. *Appl. Energy* **2011**, *88*, 68–75. [CrossRef]
16. Wang, J.Z.; Zhu, W.J.; Zhang, W.Y.; Sun, D.H. A trend fixed on firstly and seasonal adjustment model combined with the SVR for short-term forecasting of electricity demand. *Energy Policy* **2009**, *37*, 1–9. [CrossRef]
17. Kucukali, S.; Baris, K. Turkeys shorts-term gross annual electricity demand forecast by fuzzy logic approach. *Energy Policy* **2010**, *38*, 38–45. [CrossRef]
18. Nguyen, H.T.; Nabney, I.T. Short-term electricity demand and gas price forecasts using wavelet transforms and adaptive models. *Energy* **2010**, *35*, 13–21. [CrossRef]
19. Park, D.C.; Sharkawi, M.A.; Marks, R.J. Electric load forecasting using a neural network. *IEEE Trans. Power Syst.* **1991**, *6*, 442–449. [CrossRef]
20. Wang, J.Z.; Liu, F.; Song, Y.L.; Zhao, J. A novel model: Dynamic choice artificial neural network (DCANN) for an electricity price forecasting system. *Appl. Soft Comput.* **2016**, *48*, 281–297. [CrossRef]
21. Yu, F.; Xu, X.Z. A short-term load forecasting model of natural gas based on optimized genetic algorithm and improved BP neural network. *Appl. Energy* **2014**, *134*, 102–113. [CrossRef]
22. Hernandez, L.; Baladron, C.; Aguiar, J.M. Short-term load forecasting for microgrids based on artificial neural networks. *Energies* **2013**, *6*, 1385–1408. [CrossRef]
23. Pandian, S.C.; Duraiswamy, K.D.; Rajan, C.C.; Kanagaraj, N. Fuzzy approach for short term load forecasting. *Electr. Power Syst. Res.* **2006**, *76*, 541–548. [CrossRef]

24. Pai, P.F. Hybrid ellipsoidal fuzzy systems in forecasting regional electricity loads. *Energy Convers. Manag.* **2006**, *47*, 2283–2289. [CrossRef]

25. Liu, L.W.; Zong, H.J.; Zhao, E.D.; Chen, C.X.; Wang, J.Z. Can China realize its carbon emission reduction goal in 2020: From the perspective of thermal power development. *Appl. Energy* **2014**, *124*, 199–212. [CrossRef]

26. Xiao, L.; Wang, J.Z.; Dong, Y.; Wu, J. Combined forecasting models for wind energy forecasting: A case stud in China. *Renew. Sustain. Energy Rev.* **2015**, *44*, 271–288. [CrossRef]

27. Wang, J.J.; Wang, J.Z.; Li, Y.N.; Zhu, S.L.; Zhao, J. Techniques of applying wavelet de-noising into a combined model for short-term load forecasting. *Int. J. Electr. Power* **2014**, *62*, 816–824. [CrossRef]

28. Azimi, R.; Ghofrani, M.; Ghayekhloo, M. A hybrid wind power forecasting model based on data mining and wavelets analysis. *Energy Convers. Manag.* **2016**, *127*, 208–225. [CrossRef]

29. Khashei, M.; Bijari, M. An artificial neural network model for time series forecasting. *Expert Syst.* **2010**, *37*, 479–489. [CrossRef]

30. Shukur, O.B.; Lee, M.H. Daily wind speed forecasting though hybrid KF-ANN model based on ARIMA. *Renew. Energy* **2015**, *76*, 637–647. [CrossRef]

31. Niu, D.; Shi, H.; Wu, D.D. Short-term load forecasting using bayesian neural networks learned by hybrid Monte Carlo algorithm. *Appl. Soft Comput.* **2012**, *12*, 1822–1827. [CrossRef]

32. Lu, C.J.; Wang, Y.W. Combining independent component analysis and growing hierarchical self-organizing maps with support vector regression in product demand forecasting. *Int. J. Prod. Econ.* **2010**, *128*, 603–613. [CrossRef]

33. Okumus, I.; Dinler, A. Current status of wind energy forecasting and a hybrid method for hourly predictions. *Energy Convers. Manag.* **2016**, *123*, 362–371. [CrossRef]

34. Che, J.; Wang, J. Short-term electricity prices forecasting based on support vector regression and auto-regressive integrated moving average modeling. *Energy Convers. Manag.* **2010**, *51*, 1911–1917. [CrossRef]

35. Meng, A.B.; Ge, J.F.; Yin, H.; Chen, S.Z. Wind speed forecasting based on wavelet packet decomposition and artificial neural networks trained by crisscross optimization algorithm. *Energy Convers. Manag.* **2016**, *114*, 75–88. [CrossRef]

36. Elvira, L.N. Annual Electrical Peak Load Forecasting Methods with Measures of Prediction Error 2002. Ph.D. Thesis, Arizona State University, Tempe, AZ, USA, 2002.

37. Wang, J.Z.; Ma, X.L.; Wu, J.; Dong, Y. Optimization models based on GM (1,1) and seasonal fluctuation for electricity demand forecasting. *Int. J. Electr. Power* **2012**, *43*, 109–117. [CrossRef]

38. Guo, Z.H.; Wu, J.; Lu, H.Y.; Wang, J.Z. A case study on a hybrid wind speed forecasting method using BP neural network. *Knowl. Based Syst.* **2011**, *2*, 1048–1056. [CrossRef]

39. Peng, L.L.; Fan, G.F.; Huang, M.L.; Hong, W.C. Hybridizing DEMD and quantum PSO with SVR in electric load forecasting. *Energies* **2016**, *9*. [CrossRef]

40. Fan, G.F.; Peng, L.L.; Hong, W.C.; Sun, F. Electric load forecasting by the SVR model with differential empirical mode decomposition and auto regression. *Neurocomputing* **2016**, *173*, 958–970. [CrossRef]

41. Liu, H.; Tian, H.Q.; Liang, X.F.; Li, Y.F. New wind speed forecasting approaches using fast ensemble empirical model decomposition, genetic algorithm, mind evolutionary algorithm and artificial neural networks. *Renew. Energy* **2015**, *83*, 1066–1075. [CrossRef]

42. Niu, D.X.; Cao, S.H.; Zhao, L.; Zhang, W.W. *Methods for Electrical Load Forecasting and Application*; China Electric Power Press: Beijing, China, 1998.

43. Liu, L.; Wang, Q.R.; Wang, J.Z.; Liu, M. A rolling grey model optimized by particle swarm optimization in economic prediction. *Comput. Intell.* **2014**, *32*, 391–419. [CrossRef]

44. Yuan, C.; Wang, J.Z.; Tang, Y.; Yang, Y.C. An efficient approach for electrical load forecasting using distributed ART (adaptive resonance theory) & HS-ARTMAP (Hyper-spherical ARTMAP network) neural network. *Energy* **2011**, *36*, 1340–1350.

45. Hooshmand, R.A.; Amooshahi, H.; Parastegari, M. A hybrid intelligent algorithm based short-term load forecasting approach. *Int. J. Electr. Power* **2013**, *45*, 313–324. [CrossRef]

Article

Short-Term Load Forecasting Based on Wavelet Transform and Least Squares Support Vector Machine Optimized by Improved Cuckoo Search

Yi Liang [1,*], Dongxiao Niu [1], Minquan Ye [2] and Wei-Chiang Hong [3]

[1] School of Economics and Management, North China Electric Power University, Beijing 102206, China; niudx@126.com
[2] School of Economics and Management, North China Electric Power University, Baoding 071003, China; hdymq2014@163.com
[3] Department of Information Management, Oriental Institute of Technology, New Taipei 220, Taiwan; samuelsonhong@gmail.com
* Correspondence: lianglouis@126.com

Academic Editor: Sukanta Basu
Received: 31 August 2016; Accepted: 11 October 2016; Published: 17 October 2016

Abstract: Due to the electricity market deregulation and integration of renewable resources, electrical load forecasting is becoming increasingly important for the Chinese government in recent years. The electric load cannot be exactly predicted only by a single model, because the short-term electric load is disturbed by several external factors, leading to the characteristics of volatility and instability. To end this, this paper proposes a hybrid model based on wavelet transform (WT) and least squares support vector machine (LSSVM), which is optimized by an improved cuckoo search (CS). To improve the accuracy of prediction, the WT is used to eliminate the high frequency components of the previous day's load data. Additional, the Gauss disturbance is applied to the process of establishing new solutions based on CS to improve the convergence speed and search ability. Finally, the parameters of the LSSVM model are optimized by using the improved cuckoo search. According to the research outcome, the result of the implementation demonstrates that the hybrid model can be used in the short-term forecasting of the power system.

Keywords: short-term load forecasting; wavelet transform; least squares support vector machine; cuckoo search; Gauss disturbance

1. Introduction

As an important part of the management modernization of electric power systems, power load forecasting has attracted increasing attention from academics and practitioners. Power load forecasting with high precision can ease the contradiction between power supply and demand and provide a solid foundation for the stability and reliable of the power grid. However, electric load is a random non-stationary series, which is influenced by a number of factors, including economic factors, time, day, season, weather and random effects, which lead to load forecasting being a challenging subject of inquiry [1].

At present, the methods for load forecasting can be divided into two parts: classical mathematical statistical methods and approaches based on artificial intelligence. Most load forecasting theories are based on time series analysis and auto-regression models, including the vector auto-regression model (VAR) [2,3], the autoregressive moving average model (ARMA) [4–6], and so on. Time series smoothness prediction methods are criticized by researchers for their weakness of non-linear fitting capability. With the development of the electricity market, the requirement of high accuracy load forecasting is more and more strict and efficient. Therefore, artificial intelligence, which includes neural network and support vector machine, gains increasing attention by scholars. Nahi Kandil and Rene Wamkeue et al. [7] applied short-term load forecasting using the artificial neural network (ANN). The examples with real data showed the effectiveness of the proposed techniques by demonstrating that using ANN can reduce load forecasting errors, compared to various existing techniques. Feng Yu and Xiaozhong Xu [8] proposed an appropriate combinational approach, which was based on an improved back propagation neural network for short-term gas load forecasting, and the network was optimized by the real-coded genetic algorithm. D.K. Chaturvedi et al. [9] applied an algorithm that integrated wavelet transform, the adaptive genetic algorithm and a fuzzy system with a generalized neural network (GNN) to solve the short-term weekday electrical load problem. Luis Hernandez [10] presented an electric load forecast architectural model based on an ANN that performed short-term load forecasting. Nima Amjady and Farshid Keynia [11] proposed a neural network, which was optimized by a new modified harmony search technique. Pan Duan et al. [12] presented a new combined method for the short-term load forecasting of electric power systems based on the fuzzy c-means (FCM) clustering, particle swarm optimization (PSO) and support vector regression (SVR) techniques. Abdollah Kavousi-Fard, Haidar Samet and Fatemeh Marzbani [13] proposed a hybrid prediction algorithm comprised of SVR and modified firefly algorithm, and the experimental results affirmed that the proposed algorithm outperforms other techniques.

The support vector machine (SVM) [14] uses the structural risk minimization principle to convert the solution process into a convex quadratic programming problem. This overcomes some shortcomings in neural networks and has achieved a good performance in practical load forecasting [15]. The problem of hyperplane parameter selection in SVM leads to a large solving scale. In order to solve this, J.A.K. Suykens and J. Vandewalle proposed least squares support vector machine (LSSVM) as a classifier in 1999. Unlike the inequality constraints introduced in the standard SVM, LSSVM proposed equality constraints in the formulation [16]. This results in the solution being transformed from one of solving a quadratic program to a set of linear equations known as the linear Karush–Kuhn–Tucker (KKT) systems [17]. Sun Wei and Liang Yi have applied the method of LSSVM in several engineering problems, including power load forecasting [18], wind speed forecasting [19], project evaluation [20] and carbon emission prediction [21]. For example, in [18], a differential evolution algorithm-based least squares support vector regression method is proposed, and the average forecasting error is less than 1.6%, which shows better accuracy and stability than the traditional LSSVR and support vector regression. The kernel parameter and penalty factor highly effect the learning and generalization ability of LSSVM, and inappropriate parameter selection may lead to the limitation of the performance of LSSVM. However, it is possible to employ an optimization algorithm to obtain an appropriate parameter combination. The particle swarm optimization model [22] and genetic algorithm model [23] model are proposed in parameter optimization for LSSVM. In order to improve the forecasting accuracy of LSSVM, this paper applies the cuckoo search algorithm based on Gauss disturbance to optimize the parameters of LSSVM. Cuckoo search (CS) was proposed by Xin-She Yang and Suash Deb in 2009. CS is a population-based algorithm inspired by the brood parasitism of cuckoo species. It has a more efficient randomization property (with the use of Levy flight) and requires fewer parameters (population size and discovery probability only) than other optimization methods [24]. The advantage of CS is that it does not have many parameters for tuning. Evidence showed that the generated results were independent of the value of the tuning parameters. At present, the CS has been applied in many fields, such as system reliability optimization [25], optimization of biodiesel engine performance [26], load frequency control [27], solar radiation forecasting [28], and so on. In order to improve the convergence speed and the global search ability, the CS algorithm based on Gauss disturbance (GCS) is proposed in which we add Gauss perturbation to the position of the nest during the iterative process. It can increase the vitality of the change of the nest position, thus improving the convergence speed and search ability effectively.

The wavelet transform (WT) is a recently-developed mathematical tool for signal analysis [29,30]. It has been successfully applied in astronomy, data compression, signal and image processing, earthquake prediction and other fields [31]. The combination of WT and LSSVM is widely used in forecasting fields [32,33]. For example, H. Shayeghi and A. Ghasemi [33] introduce WT and improved LSSVM to predict electricity prices. The simulation results show that this technique increases electricity price market forecasting accuracy compared to the other classical and heretical methods in scientific research. Thus, this paper proposes a hybrid model based on WT and LSSVM, which is optimized by GCS, defined as W-GCS-LSSVM, and the examples demonstrate the effectiveness of the model.

The rest of the paper is organized as follows: Section 2 provides some basic theoretical aspects of WT, LSSVM and CS and gives a brief description about the W-GCS-LSSVM model; in Section 3, an experiment study is put forward to prove the efficiency of the proposed model; Section 4 is the conclusion of this paper.

2. W-GCS-LSSVM

2.1. Wavelet Transform

As an effective method for signal processing, the wavelet transform can be divided into two classifications: discrete wavelet transform (*DWT*) and continuous wavelet transform (*CWT*). The *CWT* of a signal $X(t)$ is defined as follows:

$$CWT_\psi(a,b) = \left(1/\sqrt{|a|}\right) \int_{-\infty}^{\infty} x(t) \psi^*((t-b)/a) dt \tag{1}$$

where a and b are the scale and the translation parameters, respectively. The equation applied for the *DWT* of a signal is as follows:

$$DWT_x(m,n) = (1/\sqrt{2^m}) \sum_k x_k \psi^*((k-n)/2^m) \tag{2}$$

in which m is the scale factor, $n = 1, 2...N$ is the sampling time and N is the number of samples.

As with other *WTs*, *DWT* is a kind of *WT* for which the wavelets are discretely sampled, and it captures both frequency and location information in temporal resolution; thus, *DWT* has a key advantage over Fourier transforms. In this paper, *DWT* is used in the data filtering stage.

In *WT*, a signal is similarly broken up into wavelets, which are the approximation component and detail components, in which the approximation component contains the low-frequency information (the most important part to give the signal its identity) and the detail components to reveal the flavor of the signal. Figure 1 shows a wavelet decomposition process. Firstly, the signal S is decomposed into an approximation component A1 and a detail component D1; then A1 is further decomposed into another approximation component A2 and a detail component D2 in order to meet higher level resolution; and so on, until it reaches a suitable number of levels.

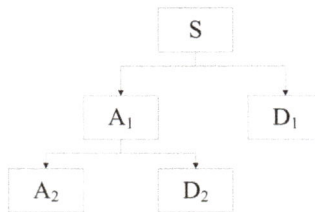

Figure 1. Wavelet decomposition.

The original short-term load data are proposed to be decomposed into one approximation component and multiple detail components. The main fluctuation of the short-term load data and the details to contain the spikes and stochastic volatilities on different levels are presented in approximation.

A suitable number of levels can be decided by comparing the similarity between the approximation and the original signal.

2.2. Least Squares Support Vector Machine

As an extension of the standard support vector machine (SVM), the least squares support vector machine (LSSVM) is proposed by SuyKens and Vandewalle [34]. By transforming the inequality constraints of traditional SVM into equality constraints, LSSVM considers the sum squares error loss function as the loss experience of the training set, which transforms solving the quadratic programming problems into solving linear equations problems [35]. The training set is set as $\{(x_k, y_k)|k = 1, 2, ..., n\}$, in which $x_k \in R^n$ and $y_k \in R^n$ represent the input data and the output data, respectively. $\phi()$ is the nonlinear mapping function, which transfers the samples into a much higher dimensional feature space $\phi(x_k)$. Establish the optimal decision function in the high-dimensional feature space:

$$y(x) = \omega^T \cdot \phi(x) + b \tag{3}$$

where $\phi(x)$ is the mapping function; ω is the weight vector; b is constant.

Using the principle of structural risk minimization, the objective optimization function is shown as follows:

$$\min_{\omega,b,e}(\omega, e) = \frac{1}{2}\omega^T \omega + \frac{1}{2}\gamma \sum_{k=1}^{n} e_k^2 \tag{4}$$

Its constraint condition is:

$$y_k = \omega^T \phi(x_k) + b + e_k \ k = 1, 2, ..., n \tag{5}$$

In which γ is the penalty coefficient and e_k represents regression error. The Lagrange method is used to solve the optimization problem; the constrained optimization problem can be transformed into an unconstrained optimization problem; the function in the dual space can be obtained as:

$$L(\omega, b, e, \alpha) = \varphi(\omega, e) - \sum_{k=1}^{n} \{\alpha_k[\omega^T \phi(x_k) + b + e_k - y_k]\} \tag{6}$$

where the Lagrange multiplier $\alpha_k \in R$. According to the Karush–Kuhn–Tucker (KKT) conditions, ω, b, e_k, α_k are taken as partial derivatives and required to be zero.

$$\begin{cases} \omega = \sum\limits_{k=1}^{n} \alpha_k \phi(x_k) \\ \sum\limits_{k=1}^{n} \alpha_k = 0 \\ \alpha_k = e_k \gamma \\ \omega^T \phi(x_k) + b + e_k - y_k = 0 \end{cases} \tag{7}$$

According to Equation (7), the optimization problem can be transformed into solving a linear problem, which is shown as follows:

$$\begin{bmatrix} 0 & 1 & \cdots & 1 \\ 1 & K(x_1, x_1) + \frac{1}{\gamma} & \cdots & K(x_1, x_l) \\ \vdots & \vdots & \vdots & \vdots \\ 1 & K(x_l, x_1) & \cdots & K(x_l, x_l) + \frac{1}{\gamma} \end{bmatrix} \begin{bmatrix} b \\ \alpha_1 \\ \vdots \\ \alpha_l \end{bmatrix} = \begin{bmatrix} 0 \\ y_1 \\ \vdots \\ y_l \end{bmatrix} \tag{8}$$

Solve Equation (8) to get α and b, then the LSSVM optimal linear regression function is:

$$f(x) = \sum_{k=1}^{l} \alpha_k K(x, x_k) + b \tag{9}$$

According to the Mercer condition, $K(x, x_i) = \phi(x)^T \cdot \phi(x_l)$ is the kernel function. In this paper, set the radial basis function (RBF) as the kernel function, which is shown in Equation (10):

$$K(x, x_k) = \exp(-\frac{|x - x_k|^2}{2\sigma^2})$$ (10)

In which σ^2 is the width of the kernel function.

From the problems of training the LSSVM, kernel parameter σ^2 and penalty parameter γ are generally set based on experience, which leads to the existence of randomness and inaccuracy in the application of the LSSVM algorithm. To solve the problem, the paper applies GCS to optimize these two parameters to improve the prediction accuracy of LSSVM.

2.3. Cuckoo Search

The cuckoo search (CS) algorithm is a new optimization metaheuristic algorithm [24], which is on the basis of the stochastic global search and the obligate brood-parasitic behavior of cuckoos by laying their eggs in the nests of host birds. In this optimization algorithm, each nest represents a potential solution. The cuckoo birds choose recently-spawned nests, so that they can be sure that eggs could hatch first because a cuckoo egg usually hatches earlier than its host bird. In addition, by mimicking the host chicks, a cuckoo chick can deceive the host bird to grab more food resources. If the host birds discover that an alien cuckoo egg has been laid (with the probability p_a), they either propel the egg or abandon the nest and completely build a new nest in a new location. New eggs (solutions) laid by the cuckoo choose the nest by Levy flights around the current best solutions. Additionally, with the Levy flight behavior, the cuckoo speeds up the local search efficiency.

Yang and Deb simplified the cuckoo parasitic breeding process by the following three idealized rules [24]:

(i)   Each cuckoo lays only one egg at a time and randomly searches for a nest in which to lay it.
(ii) An egg of high quality will be considered to survive to the next generation.
(iii) The number of available host nests is fixed, and a host can discover an alien egg with a probability $p_a \in [0, 1]$. In this case, the host bird can either throw the egg away or abandon the nest so as to build a completely new nest in a new location. The last strategy is approximated by a fraction p_a of the n nests being replaced by new nests (with new random solutions at new locations).

In sum, two search capabilities have been used in cuckoo search: global search (diversification) and local search (intensification), controlled by a switching/discovery probability (p_a). Local search can be described as follows:

$$x_i^{(t+1)} = x_i^t + \alpha s \oplus H(p_a - \varepsilon) \otimes (x_j^t - x_k^t)$$ (11)

where x_j^t and x_k^t are different random sequences; $H(u)$ is the Hedwig–Cede function; ε represents a random number; s means the step lengths. The global search is based on Levy flight, which is shown as follows:

$$x_i^{(t+1)} = x_i^t + \alpha \oplus L(s, \lambda)$$ (12)

where $L(s, \lambda) = \frac{\lambda \Gamma(\lambda) \sin(\pi \lambda / 2)}{\pi} \frac{1}{s^{1+\lambda}}$, $s >> s_0$, $1 < \lambda \leq 3$; α is the levy flight step size multiplication processes with an entry-wise multiplication process. The product $\oplus$ means entry-wise multiplications, which is similar to those used in PSO, but the random walk process via Levy flight here is more efficient in exploring the search space, for its step length is much longer in the long run. It is worth pointing out that, in the real world, if a cuckoo's egg is very similar to a host's eggs, then this cuckoo's egg is less likely to be discovered; thus, the fitness should be related to the difference in solutions. Therefore, it is a good idea to do a random walk in a biased way with some random step size location [36].

The pseudo-code for the CS is performed in Figure 2:

```
Objective Function f(x),x=(x₁,x₂, ···,xₐ)
Generate initial population of n host nest xᵢ(i−1,2, ···,n)
While( t < MaxGeneration)
  Get a cuckoo randomly by Levy flights
  Evaluate its quality/fitness Fᵢ
  Chose a nest among n (say, j) randomly
  If( Fᵢ > Fⱼ)
  Replace j by the new solution
  end
A fraction (pₐ) of worst nests are abandoned and new ones are built;
Keep the solutions(or nests with quality solutions)
Rank the solutions and find the current best
end While
```

Figure 2. The pseudo-code of the cuckoo search (CS).

2.4. CS Algorithm Based on Gauss Disturbance

On the basis of the cuckoo search algorithm (CS), the CS algorithm based on the Gauss disturbance (GCS) is proposed in which we add Gauss perturbation to the position of the nest during the iterative process. It can increase the vitality of the change of the nest position, thus improving the convergence speed and search ability effectively.

The basic idea of the cuckoo search algorithm (CS) based on Gauss perturbation is: continue to conduct the Gauss perturbation of x_i^t to make a further search instead of coming directly into the next iteration when a better set of nest locations $x_i^t, i = 1, 2, ..., n$ is gained after t iterations in CS. Suppose $x_i^t, i = 1, 2, ..., n$ is a d-dimensional vector and p_t is described as $p_t = [x_1^{(t)}, x_2^{(t)}, ..., x_n^{(t)}]^T$, then p_t is a $d \times n$ matrix. The specific operation of GCS algorithm is adding Gauss perturbation to p_t, namely:

$$p_t' = p_t + a \oplus \varepsilon \tag{13}$$

where ε is a random matrix of the same order to p_t, $\varepsilon_{ij} \sim N(0, 1)$; a is a constant; $\oplus$ represents the point-to-point multiplication. The large range of the value of ε easily leads to the large deviation of the nest location. Therefore, we select $a = 1/3$ to control the search scope of ε, thus moderately increasing the vitality of the change of the nest position to make p_t' reasonable. Then, compare it with each nest in p_t and update p_t'' with a better set of nest positions. For the next iteration, p_t'' can be represented as $p_t = [x_1^{(t)}, x_2^{(t)}, ..., x_n^{(t)}]^T$.

2.5. LSSVM Optimized by the CS Algorithm Based on Gauss Disturbance

The flowchart of the W-GCS-LSSVM model is shown in Figure 3, and the detailed processes are as follows:

(1) Decompose the load signal into the approximation A1 and the details D1, and select A1 as the training data and testing data. Normalize the load data.

(2) Determine the value range of σ^2 and γ of LSSVM and related parameters of GCS. In this paper, the number of host nests is 25; the maximum number of iterations is 400; and the search range is between 0.01 and 100.

(3) Suppose the initial probability parameter p_a is 0.25, and set $p_i^0 = [x_1^0, x_2^0, ..., x_n^0]^T$ as the location of a random n nest. Each nest corresponds to a set of parameters (σ^2, γ). Then, calculate the fitting degree of each nest position to find the best nest location x_b^0 and the minimum fitting degree $F_{\min}$. The root mean square error (*RMSE*) is applied as the fitness function:

$$RMSE = \sqrt{\frac{\sum_{i=1}^{n} (y_i - \hat{y}_i)^2}{n}} \tag{14}$$

(4) Reserve the best nest position x_b^0 and update other nest positions through Levy flights to obtain a new set of nest positions, then calculate the fitting degree F.

(5) Compare the new nest positions with the preceding generation p_{i-1} according to the fitting degree F and update the nest position with a better one; thus, the new set nest position is described as follows: $p_t = [x_1^t, x_2^t, ..., x_n^t]^T$.

(6) Compare the p_a to a random number r. Reserve the nests with lower probability to be discovered in p_t and replace the higher one. Then, calculate the fitting degree of the new nests and update the nest position p_t by comparing it with the precedent fitness degree.

(7) Obtain a new set of nest positions $p_t' = [x_1^t, x_2^t, ..., x_n^t]^T$ through Gaussian perturbation of p_t. Then, compare the test value of p_t' with p_t. Update the nest positions with better test values as $p_t'' = [x_1^t, x_2^t, ..., x_n^t]^T$. Here, p_t'' is denoted by $p_t = [x_1^{(t)}, x_2^{(t)}, ..., x_n^{(t)}]^T$ for the next iteration.

(8) Find the best nest position x_b^t in Step (7). If the fitting degree F meets the requirements, stop the algorithm, and then, output the global minimum fitting degree F_{min}, as well as the best nest x_b^t. If not, return to Step (4) to continue optimization.

(9) Set the optimal parameters σ^2 and γ of LSSVM according to the best nest position x_b^t.

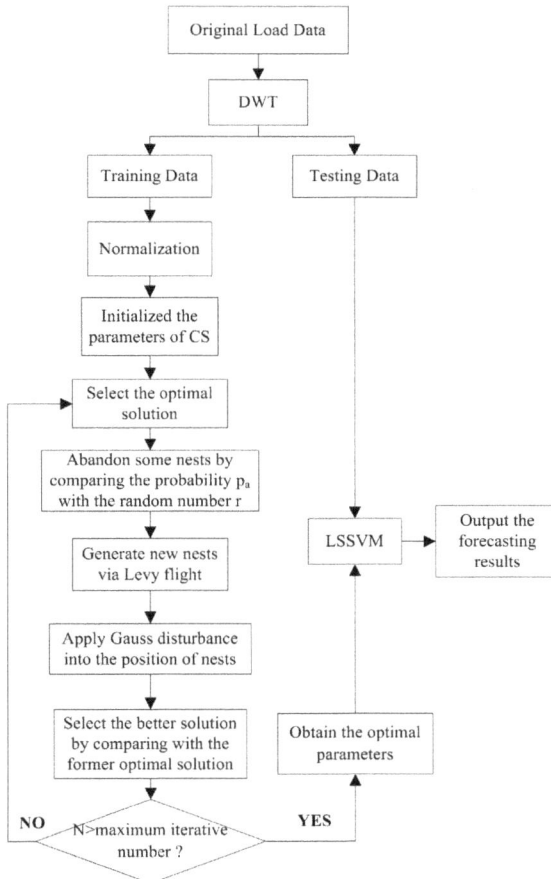

Figure 3. Flowchart of the W-CS algorithm based on Gauss disturbance (GCS)-least squares support vector machine (LSSVM) modeling.

3. Case Study

3.1. Data Preprocessing

This paper establishes a prediction model of short-term load forecasting and analyzes the prediction results of the examples. The 24-h short-term load forecasting has been made on the power system of Yangquan city in China from 1 April to 30 May 2013 (the load data of 23 May are missing). Figure 4 shows the power load of 1416 samples, ranging from around 730 MW to 950 MW. From Figure 4, no apparent regularity of power load can be obtained. In this paper, we select 708 load data from 1 to 30 April as the training set, 660 load data from 30 April to 28 May as the validation set and 72 load data from 29 to 31 May as the testing set.

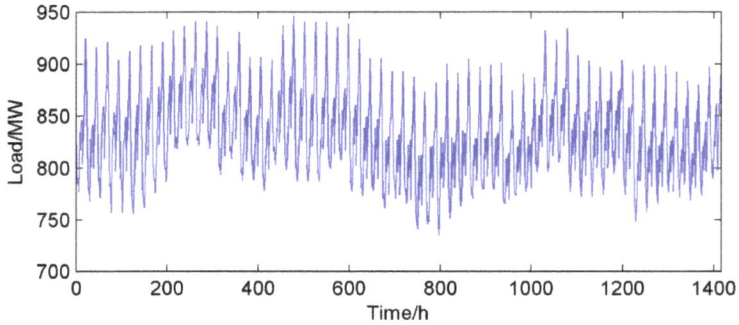

Figure 4. Load curve for each hour.

The original load data are decomposed to eliminate the current precipitation value for further modeling by using *WT*. The original short-term load data S and their approximation A1, as well as the detail component D1 decomposed by one-level *DWT* are shown in Figure 5.

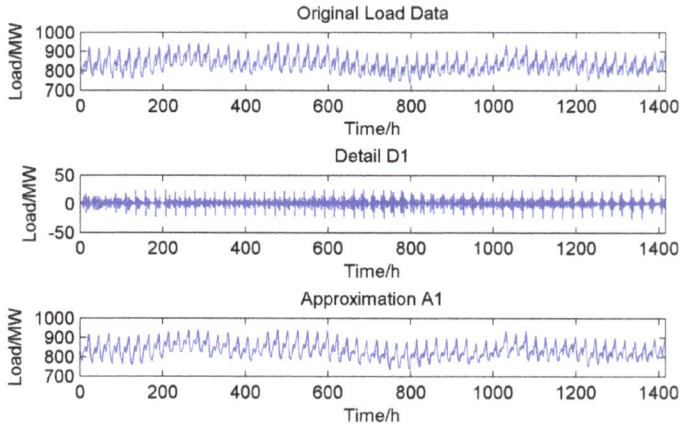

Figure 5. Original load signal and its approximation component and detail component decomposed by *DWT*.

From Figure 5, it can be clearly seen that A1, which presents the major fluctuation of the original short-term load data, shows high similarity to S; meanwhile, the other minor irregularity neglected by A1 appears in D1. Therefore, A1 is taken as the input data to model for efficiency.

3.2. Selection of Input

Human activities are always disturbed by many external factors, and then, the power load is affected. Therefore, some effective features are considered as input features. In this paper, the input features are discussed as follows. (1) The temperature: Temperature is one of these effective features. In previous studies [37,38], temperature was considered as an essential input feature and the forecasting results were accurate enough. The curves of temperature and load data are shown in Figure 6. Therefore, the temperature is taken into consideration. (2) Weather conditions: The weather conditions are divided into four types: sunny, cloudy, overcast and rainy. For different weather conditions, we set different weights: {sunny, cloudy, overcast, rainy} = {0.8, 0.6, 0.4, 0.2}. (3) Day type: For different day types, the electric power consumption is different. Figure 7 shows the load data from 28 April to 4 May 2013. From Figure 7, we can see that different day types have different curve features. Therefore, we assign values to the day type in Table 1.

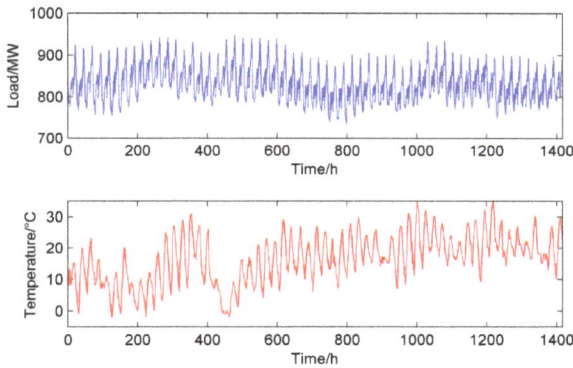

Figure 6. The curves of the load data and temperature.

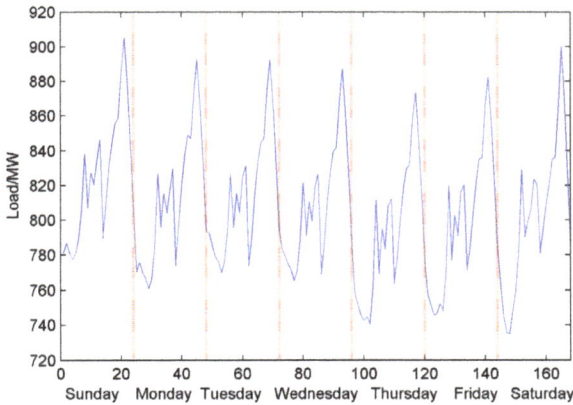

Figure 7. Weekly load curve.

Table 1. The values of the day type.

Day Type	Monday	Tuesday	Wednesday	Thursday	Friday	Saturday	Sunday
Weights	1	2	3	4	5	6	7

3.3. Model Performance Evaluation

The work in [39] discusses and compares the measures of the accuracy of univariate time series forecasts. According to this reference, the relative error (RE), the mean absolute percentage error (MAPE), the root mean square error (*RMSE*) and absolute error (AE) are proposed to measure the forecast accuracy. The equations are as follows:

$$RE(i) = \frac{\hat{y}_i - y_i}{y_i} \times 100\% \tag{15}$$

$$MAPE = \frac{1}{n} \sum_{i=1}^{n} \left| \frac{\hat{y}_i - y_i}{y_i} \right| \tag{16}$$

$$MSE = \frac{1}{n} \sum_{i=1}^{n} (\hat{y}_i - y_i)^2 \tag{17}$$

$$AE = \left| \frac{\hat{y}_i - y_i}{y_i} \times 100\% \right| \tag{18}$$

where y_i represents the actual value at period i; $\hat{y}_i$ is the forecasting value at period i; and n is the number of forecasting period.

3.4. Analysis of Forecasting Results

At first, the GCS is used to optimize the kernel parameter σ^2 and penalty parameter γ in LSSVM. The parameter settings of GCS is given in Section 2.4. Figure 8 shows the iterations process of GCS. From the figure we can see that GCS achieves convergence at 263 times. The optimal values of σ^2 and γ are respectively 6.41 and 16.24.

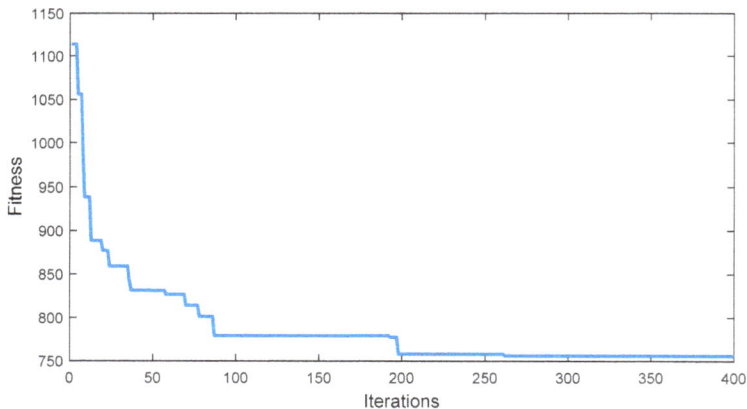

Figure 8. The iterations process of GCS.

The short-term electric load forecasting results of three days of the W-GCS-LSSVM, GCS-LSSVM, CS-LSSVM, W-LSSVM ($\sigma^2 = 5$ and $\gamma = 10$) and LSSVM ($\sigma^2 = 5$ and $\gamma = 10$) model are respectively shown in Tables 2–4. In order to explain the results more clearly, the proposed model and comparison models are divided into two groups: the first group includes W-GCS-LSSVM, GCS-LSSVM and CS-LSSVM, and the second group consists of W-GCS-LSSVM, W-LSSVM and LSSVM, which are respectively shown in Figures 9 and 10. Moreover, Figures 11 and 12 show the comparisons of relative errors between the proposed model and the others. The RE ranges [−3%, 3%] and [−1%, 1%] are popularly regarded as a standard to evaluate the performance of a prediction model [40]. Based on these tables and figures, we can obtain that: (1) the REs of the short-term load forecasting model of W-GCS-LSSVM are all in the range of [−3%, 3%]; the maximum RE is 2.4380% at 15:00 (Day 1), and the minimum RE is −2.901% at 13:00 (Day 1); there exists thirty points that are in the scope of [−1%, 1%]; (2) the GCS-LSSVM has three predicted points that exceed the RE range [−3%, 3%], which are 4.5000% at 19:00 (Day 3), 3.6472% at 8:00 (Day 3) and 3.1826% at 16:00 (Day 1), and there are twenty-nine predicted points in the range of [−1%, 1%]; (3) the CS-LSSVM has four predicted points that exceed the RE range [−3%, 3%], which are 5.0824% at 19:00 (Day 3), 3.1026% at 22:00 (Day 2), 3.0863% at 16:00 (Day 1) and −3.0154% at 17:00 (Day 2), and there are twenty-one predicted points in the range of [−1%, 1%]; (4) the W-LSSVM has four predicted points that exceed the RE range [−3%, 3%], which are respectively 3.4763% at 17:00 (Day 1), 3.2786% at 0:00 (Day 2), 3.0215% at 11:00 (Day 2) and −3.3465% at 13:00 (Day 1), and there are seventeen predicted points in the range of [−1%, 1%]; (5) the single LSSVM has fourteen predicted points that exceed the RE range [−3%, 3%], which are respectively 4.1518% at 8:00 (Day 1), 3.8082% at 23:00 (Day 1), 3.4807% at 12:00 (Day 3), 3.4028% at 23:00 (Day 2), 3.3572% at 16:00 (Day 3), 3.3091% at 17:00 (Day 1), 3.2287% at 17:00 (Day 3), 3.1997% at 7:00 (Day 1), 3.1958% at 15:00 (Day 3), 3.0350% at 13:00 (Day 3), −3.1991% at 5:00 (Day 1), −3.2325% at 3:00 (Day 2) and −3.5397% at 14:00 (Day 1), and there are fifteen predicted points in the range of [−1%, 1%]. From the global view of RE, the forecasting accuracy of W-GCS-LSSVM is better than the other models, since it has the most predicted points in the ranges [−1%, 1%] and [−3%, 3%]. Moreover, from Figure 9, the results of GCS-LSSVM are better than those of CS-LSSVM, which can verify that the Gauss disturbance strategy applied in CS increases the vitality of the change of the nest position, thus improving the convergence speed and search ability effectively. From Figure 10, the effects of W-LSSVM are better than single LSSVM, which can illustrate that *WT* effectively filters the original data. However, the comparison models also predict more accurately than the proposed model at some points, for example the RE of W-GCS-LSSVM is 2.901% at 13:00 (Day 1), which is higher than that of GCS-LSSVM, CS-LSSVM and LSSVM.

Table 2. Actual load and forecasting results in Day 1 (Unit: MV).

Time/h	Actual Data	W-GCS-LSSVM	GCS-LSSVM	CS-LSSVM	W-LSSVM	LSSVM
D1 0:00	819.22	824.19	816.24	824.33	818.62	808.19
D1 1:00	794.17	795.39	791.85	793.01	794.81	792.45
D1 2:00	781	780.58	788.85	789.83	779.74	798.44
D1 3:00	774.72	777.43	786.75	782.96	785.82	778.57
D1 4:00	772.77	778.34	779.29	781.87	782.15	788.59
D1 5:00	782.96	770.16	775.57	773.96	771.73	757.91
D1 6:00	788.06	784.95	784.70	783.63	785.81	784.52
D1 7:00	805.28	815.59	818.14	815.14	813.85	831.05
D1 8:00	814.13	821.34	829.90	830.01	827.35	847.93
D1 9:00	804.14	811.42	795.91	791.91	817.41	809.03
D1 10:00	822.51	813.97	816.38	813.35	816.82	811.04
D1 11:00	831.4	833.61	819.48	820.36	833.63	814.20
D1 12:00	844.94	835.04	837.60	836.71	834.76	830.21
D1 13:00	849.24	824.61	844.59	845.87	820.82	866.12
D1 14:00	804.53	819.21	796.47	796.34	818.67	776.05
D1 15:00	791.98	811.29	810.95	810.87	811.38	810.86
D1 16:00	802.18	818.43	827.71	826.94	818.47	838.65
D1 17:00	816.86	835.09	840.89	840.88	845.26	843.89
D1 18:00	837.08	855.90	857.95	858.30	855.93	857.99
D1 19:00	852.35	853.37	853.06	853.28	853.40	869.15
D1 20:00	856.64	869.55	864.12	865.014	867.90	836.73
D1 21:00	880.66	900.40	903.09	903.91	899.78	902.18
D1 22:00	881	897.83	889.43	895.33	893.81	898.87
D1 23:00	833.55	845.33	848.99	850.65	845.25	865.29

Table 3. Actual load and forecasting results in Day 2 (Unit: MV).

Time/h	Actual Data	W-GCS-LSSVM	GCS-LSSVM	CS-LSSVM	W-LSSVM	LSSVM
D2 0:00	820.46	832.43	843.35	836.78	847.36	838.42
D2 1:00	805.16	812.32	816.74	825.76	814.57	818.54
D2 2:00	798.03	782.32	785.59	807.47	793.76	780.42
D2 3:00	799.06	804.94	812.42	819.58	815.87	773.23
D2 4:00	805.05	813.26	805.56	801.75	808.95	815.53
D2 5:00	805.42	810.52	798.67	792.84	822.46	815.34
D2 6:00	820.92	809.91	814.75	811.86	812.87	829.43
D2 7:00	841.42	832.62	849.53	859.54	821.74	832.58
D2 8:00	824.37	837.73	813.65	804.93	812.56	845.76
D2 9:00	846.60	863.42	868.87	857.75	842.43	832.43
D2 10:00	860.55	864.53	853.67	858.82	868.52	872.54
D2 11:00	867.44	887.29	882.56	875.26	893.65	851.76
D2 12:00	863.01	872.42	846.64	853.57	865.78	867.34
D2 13:00	817.65	809.64	803.56	835.53	826.68	825.86
D2 14:00	818.51	813.93	832.67	826.82	822.56	802.65
D2 15:00	839.02	836.22	852.57	863.98	824.75	823.75
D2 16:00	858.49	873.12	864.67	861.79	882.79	864.25
D2 17:00	879.16	874.64	862.76	852.65	877.53	885.29
D2 18:00	902.11	915.82	894.73	906.64	907.75	924.63
D2 19:00	884.54	903.54	908.47	892.88	908.64	899.43
D2 20:00	917.62	916.37	937.43	927.45	927.65	934.54
D2 21:00	919.17	930.48	912.57	925.75	937.73	902.43
D2 22:00	890.22	901.54	899.73	917.84	906.43	914.35
D2 23:00	843.72	852.45	832.76	826.87	862.58	872.43

Table 4. Actual load and forecasting results in Day 3 (Unit: MV).

Time/h	Actual Data	W-GCS-LSSVM	GCS-LSSVM	CS-LSSVM	W-LSSVM	LSSVM
D3 0:00	799.38	783.43	808.87	802.43	787.86	813.50
D3 1:00	784.48	792.66	789.43	794.62	801.54	806.64
D3 2:00	777.53	784.34	768.59	759.52	781.48	785.74
D3 3:00	778.53	787.23	779.76	783.59	793.78	782.74
D3 4:00	784.36	802.98	779.25	775.24	794.65	805.99
D3 5:00	784.72	796.32	790.31	778.98	804.92	790.22
D3 6:00	799.83	792.23	806.98	812.76	787.77	811.39
D3 7:00	819.81	813.87	815.42	811.46	826.41	819.27
D3 8:00	808.02	812.59	837.49	822.54	802.83	812.74
D3 9:00	829.81	837.31	848.26	844.72	823.75	836.71
D3 10:00	843.49	832.98	849.72	837.28	845.48	850.32
D3 11:00	855.36	862.48	866.74	870.62	867.74	877.88
D3 12:00	850.99	857.55	841.53	836.66	852.65	880.61
D3 13:00	806.26	813.69	805.87	800.43	825.98	830.73
D3 14:00	807.11	819.43	816.76	804.58	814.65	825.97
D3 15:00	827.34	814.87	812.83	836.65	810.54	853.78
D3 16:00	846.53	837.49	849.23	855.92	823.65	874.95
D3 17:00	866.92	874.43	871.59	864.46	863.42	894.91
D3 18:00	889.56	897.78	902.57	909.34	902.67	908.75
D3 19:00	872.23	893.45	911.48	916.56	897.85	877.78
D3 20:00	904.85	916.77	893.56	887.94	924.43	898.62
D3 21:00	906.38	909.49	917.34	922.54	916.49	926.14
D3 22:00	867.61	882.73	892.52	885.91	877.61	853.82
D3 23:00	831.98	841.76	845.46	847.43	835.64	856.50

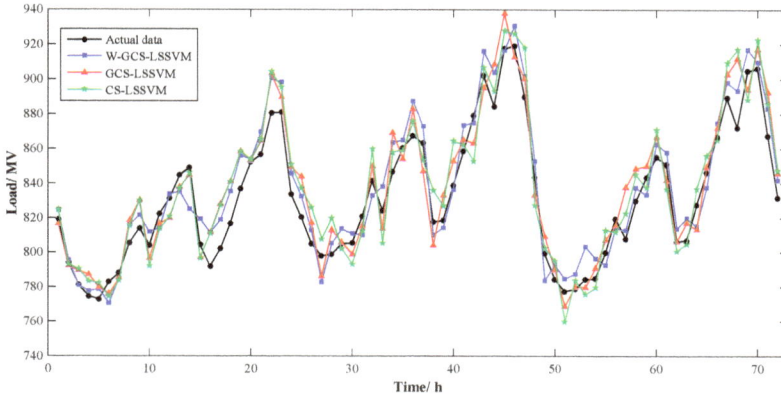

Figure 9. Actual load and forecasting results of W-GCS-LSSVM, GCS-LSSVM and CS-LSSVM.

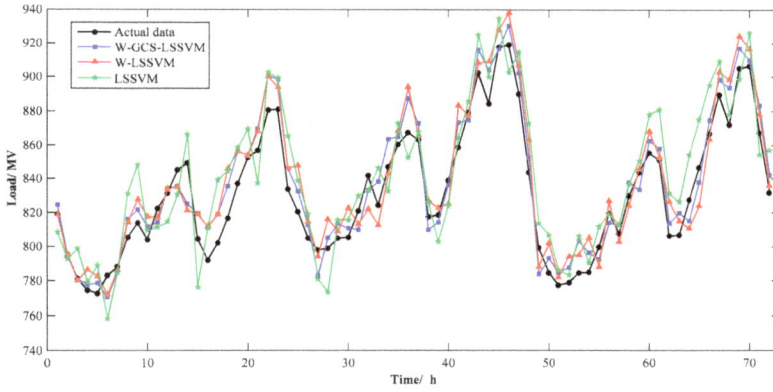

Figure 10. Actual load and forecasting results of W-GCS-LSSVM, W-LSSVM and LSSVM.

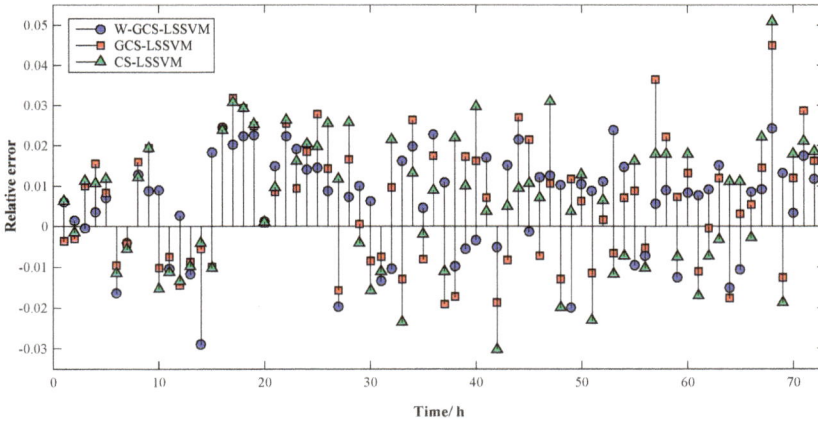

Figure 11. Relative Errors of W-GCS-LSSVM, GCS-LSSVM and CS-LSSVM.

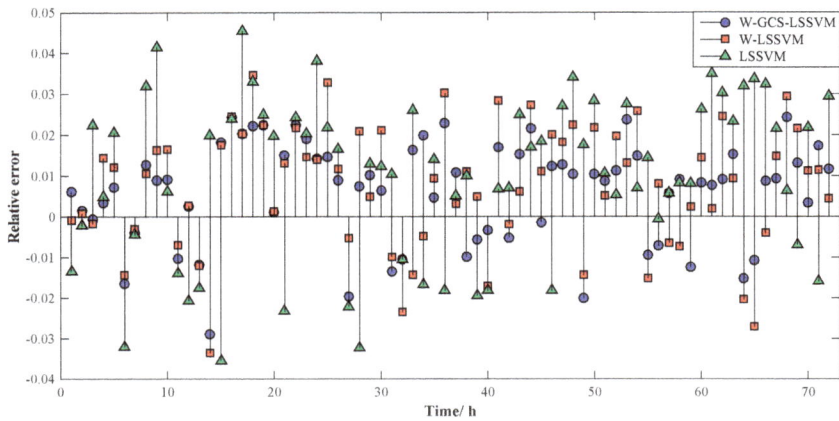

Figure 12. Relative Errors of W-GCS-LSSVM, W-LSSVM and LSSVM.

The MAPE and MSE of WT-GCS-LSSVM, GCS-LSSVM, CS-LSSVM, W-LSSVM and LSSVM are listed in Table 5. From Table 5, we can conclude that the MAPE of the proposed model is 1.2083%, which is smaller than the MAPE of GCS-LSSVM, CS-LSSVM, W-LSSVM and LSSVM (which are 1.3682%, 1.4790%, 1.4213% and 1.9557%). In addition, the MSE of the proposed model is 131.6950, which is smaller than the MSE of the comparison models (which are 185.6538, 210.7736, 196.6906 and 336.5224). As a result, the MAPE and MSE of the W-GCS-LSSVM are both smaller than those of the W-LSSVM, so we can conclude that the parameter optimization to LSSVM is essential in the forecasting model. Besides, the MAPE and MSE of the WT-GCS-LSSVM are both smaller than GCS-LSSVM, indicating the pre-processing of load data is useful for a better performance and higher forecasting accuracy. At the same time, the MAPE and MSE of the GCS-LSSVM and CS-LSSVM are both smaller than those of LSSVM, and this presents that the optimization results of the GCS and CS are efficient.

Table 5. Model performance evaluations.

Model	W-GCS-LSSVM	GCS-LSSVM	CS-LSSVM	W-LSSVM	LSSVM
MAPE	1.2083%	1.3682%	1.4790%	1.4213%	1.9557%
MSE	131.6950	185.6538	210.7736	196.6906	336.5224

In addition, the AE of the load forecasting value divided into four parts that is calculated from Equation (18) is shown in Figure 13. The numbers on the x-axis represent the models appeared above: 1 represents the W-GCS-LSSVM model, 2 represents the GCS-LSSVM model, 3 represents the CS-LSSVM model, 4 represents the W-LSSVM model and 5 represents the single LSSVM model. From Figure 13, we can discover that the AE values of W-GCS-LSSVM are almost lower than those of the other models. The numbers of points that are less than 1%, 3% and more than 3% and the corresponding percentage of them in the predicted points are accounted, respectively. The statistical results are shown in Table 6. It can be seen that there are 30 predicted points whose the AE of the W-GCS-LSSVM model is less than 1%, which accounts for 41.67% of the total amount; and 42 predicted points less than 3%, accounting for 58.33% of the total amount. Besides, there are no number predicted points whose AE is more than 3%, accounting for 0% of the total amount. It can be indicated that the prediction performance of the proposed model is superior, and its accuracy is higher. Therefore, the W-GCS-LSSVM model is suitable for short-term load forecasting.

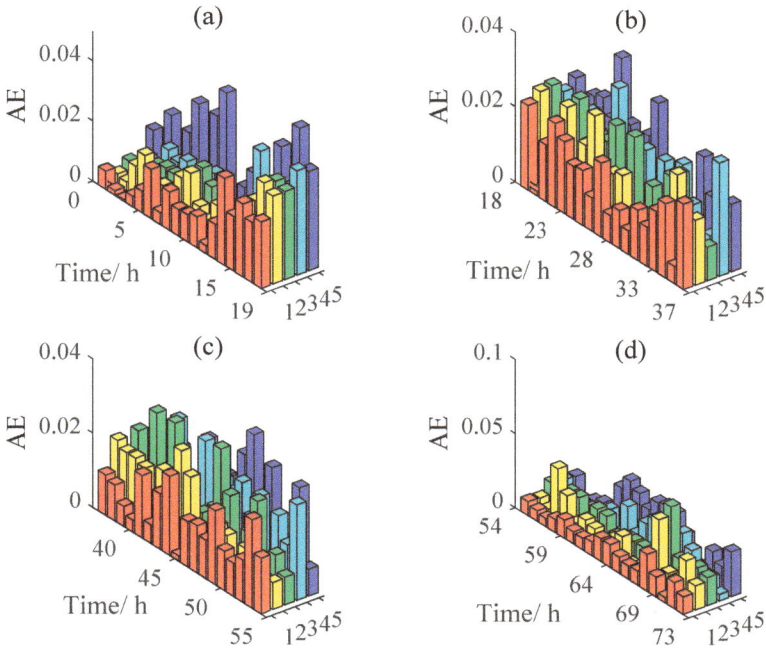

Figure 13. Absolute error distribution curve for different models, (**a**) the AE value from 1 to 18 sample point; (**b**) the AE value from 19 to 36 sample point; (**c**) the AE value from 37 to 54 sample point; (**d**) the AE value from 55 to 72 sample point.

Table 6. Accuracy estimation of the prediction point for the test set.

Prediction Model	<1%		>1% and <3%		≥3%	
	Number	Percentage	Number	Percentage	Number	Percentage
W-GCS-LSSVM	30	41.67%	42	58.33%	0	0%
GCS-LSSVM	29	40.28%	40	55.56%	3	4.17%
CS-LSSVM	21	29.17%	47	65.28%	4	5.56%
W-LSSVM	25	34.72%	43	59.72%	4	5.56%
LSSVM	15	20.83%	43	59.72%	14	19.44%

4. Conclusions

To strengthen the stability and economy of the grid and avoid waste in grid scheduling, it is essential to improve the forecasting accuracy. Because the short-term power load is always interfered with by various external factors with characteristics like high volatility and instability, the high accuracy of load forecasting should be taken into consideration. Based on the features of load data and the randomness of the LSSVM parameter setting, we propose the model based on wavelet transform and least squares support vector machine optimized by improved cuckoo search. To validate the proposed model, four other comparison models (GCS-LSSVM, CS-LSSVM, W-LSSVM and LSSVM) are employed to compare the forecasting results. Example computation results show that the relative errors of the W-GCS-LSSVM model are all in the range of [−3%, 3%], and the MAPE and MSE are both smaller than the others. In addition, the advantage of CS is that it does not have many parameters for tuning, so it can be applied widely in parameter optimization. However, seasonality and long-term trend of the proposed model are not tested and verified in this paper, which may become the limitation of this method, and authors will study it in the future. Above all, the hybrid model can be effectively used in the short-term load forecasting on the power system.

Acknowledgments: This work is supported by the Natural Science Foundation of China (Project No. 71471059).

Author Contributions: Yi Liang designed this research and wrote this paper; Dongxiao Niu and Wei-Chiang Hong provided professional guidance; and Minquan Ye collected all the data and revised this paper.

Conflicts of Interest: The authors declare no conflict of interest.

References

1. Bozic, M.; Stojanovic, M.; Stajic, Z.; Tasić, D. A new two-stage approach to short term electrical load forecasting. *Energies* **2013**, *6*, 2130–2148. [CrossRef]
2. Liang, R.H.; Chen, Y.K.; Chen, Y.T. Volt/Var control in a distribution system by a fuzzy optimization approach. *Int. J. Electr. Power Energy Syst.* **2011**, *33*, 278–287. [CrossRef]
3. Niknam, T.; Zare, M.; Aghaei, J. Scenario-based multiobjective Volt/Var control in distribution networks including renewable energy sources. *IEEE Trans. Power Deliv.* **2012**, *27*, 2004–2019. [CrossRef]
4. Lee, C.M.; Ko, C.N. Short-term load forecasting using lifting scheme and ARIMA models. *Expert Syst. Appl.* **2011**, *38*, 5902–5911. [CrossRef]
5. Pappas, S.S.; Ekonomou, L.; Karamousantas, D.C.; Chatzarakis, G.E.; Katsikas, S.K.; Liatsis, P. Electricity demand loads modeling using Auto Regressive Moving Average (ARMA) models. *Energy* **2008**, *33*, 1353–1360. [CrossRef]
6. Pappas, S.S.; Ekonomou, L.; Karampelas, P.; Karamousantasd, D.C.; Katsikase, S.K.; Chatzarakis, G.E.; Skafidas, P.D. Electricity demand load forecasting of the Hellenic power system using an ARMA model. *Electr. Power Syst. Res.* **2010**, *80*, 256–264. [CrossRef]
7. Kandil, N.; Wamkeue, R.; Saad, M.; Georges, S. An efficient approach for short term load forecasting using artificial neural networks. *Int. J. Electr. Power Energy Syst.* **2006**, *28*, 525–530. [CrossRef]
8. Yu, F.; Xu, X. A short-term load forecasting model of natural gas based on optimized genetic algorithm and improved BP neural network. *Appl. Energy* **2014**, *134*, 102–113. [CrossRef]
9. Chaturvedi, D.K.; Sinha, A.P.; Malik, O.P. Short term load forecast using fuzzy logic and wavelet transform integrated generalized neural network. *Int. J. Electr. Power Energy Syst.* **2015**, *67*, 230–237. [CrossRef]
10. Hernandez, L.; Baladrón, C.; Aguiar, J.M.; Carro, B.; Sanchez-Esguevillas, A.J.; Lloret, J. Short-Term Load Forecasting for Microgrids Based on Artificial Neural Networks. *Energies* **2013**, *6*, 1385–1408. [CrossRef]
11. Amjady, N.; Keynia, F. A new neural network approach to short term load forecasting of electrical power systems. *Energies* **2011**, *4*, 488–503. [CrossRef]
12. Pan, D.; Xie, K.; Guo, T.; Huang, X. Short-term load forecasting for electric power systems using the PSO-SVR and FCM clustering techniques. *Energies* **2011**, *4*, 173–184.
13. Kavousi-Fard, A.; Samet, H.; Marzbani, F. A new hybrid Modified Firefly Algorithm and Support Vector Regression model for accurate Short Term Load Forecasting. *Expert Syst. Appl.* **2014**, *41*, 6047–6056. [CrossRef]
14. Vapnik, V. *The Nature of Statistical Learning Theory*; Springer Science & Business Media: Berlin, Germany, 2000.
15. Sun, W.; Liang, Y. Least-Squares Support Vector Machine Based on Improved Imperialist Competitive Algorithm in a Short-Term Load Forecasting Model. *J. Energy Eng.* **2014**, *141*, 04014037. [CrossRef]
16. Mesbah, M.; Soroush, E.; Azari, V.; Lee, M.; Bahadori, A.; Habibnia, S. Vapor liquid equilibrium prediction of carbon dioxide and hydrocarbon systems using LSSVM algorithm. *J. Supercrit. Fluids* **2015**, *97*, 256–267. [CrossRef]
17. Liu, H.; Yao, X.; Zhang, R.; Liu, M.; Hu, Z.; Fan, B. Accurate quantitative structure-property relationship model to predict the solubility of C60 in various solvents based on a novel approach using a least-squares support vector machine. *J. Phys. Chem. B* **2005**, *109*, 20565–20571. [CrossRef] [PubMed]
18. Sun, W.; Liang, Y. Research of least squares support vector regression based on differential evolution algorithm in short-term load forecasting model. *J. Renew. Sustain. Energy* **2014**, *6*, 053137. [CrossRef]
19. Sun, W.; Liu, M.H.; Liang, Y. Wind speed forecasting based on FEEMD and LSSVM optimized by the bat algorithm. *Energies* **2015**, *8*, 6585–6607. [CrossRef]

20. Sun, W.; Liang, Y. Comprehensive evaluation of cleaner production in thermal power plants using particle swarm optimization based least squares support vector machines. *J. Inf. Comput. Sci.* **2015**, *12*, 1993–2000. [CrossRef]

21. Sun, W.; Liang, Y.; Xu, Y.F. Application of carbon emissions prediction using least squares support vector machine based on grid search. *WSEAS Trans. Syst. Control* **2015**, *10*, 95–104.

22. Gorjaei, R.G.; Songolzadeh, R.; Torkaman, M.; Safari, M.; Zargar, G. A novel PSO-LSSVM model for predicting liquid rate of two phase flow through wellhead chokes. *J. Nat. Gas Sci. Eng.* **2015**, *24*, 228–237. [CrossRef]

23. Liu, D.; Niu, D.; Wang, H.; Fan, L. Short-term wind speed forecasting using wavelet transform and support vector machines optimized by genetic algorithm. *Renew. Energy* **2014**, *62*, 592–597. [CrossRef]

24. Yang, X.S.; Deb, S. Cuckoo Search via Levy Flights. In Proceedings of the World Congress on IEEE 2010, Barcelona, Spain, 18–23 July 2010.

25. Valian, E.; Tavakoli, S.; Mohanna, S.; Haghi, A. Improved cuckoo search for reliability optimization problems. *Comput. Ind. Eng.* **2013**, *64*, 459–468. [CrossRef]

26. Wong, P.K.; Wong, K.I.; Chi, M.V.; Cheung, C.S. Modelling and optimization of biodiesel engine performance using kernel-based extreme learning machine and cuckoo search. *Renew. Energy* **2015**, *74*, 640–647. [CrossRef]

27. Abdelaziz, A.Y.; Ali, E.S. Cuckoo Search algorithm based load frequency controller design for nonlinear interconnected power system. *Int. J. Electr. Power Energy Syst.* **2015**, *73*, 632–643. [CrossRef]

28. Wang, J.; Jiang, H.; Wu, Y.; Dong, Y. Forecasting solar radiation using an optimized hybrid model by Cuckoo Search algorithm. *Energy* **2015**, *81*, 627–644. [CrossRef]

29. Deihimi, A.; Orang, O.; Showkati, H. Short-term electric load and temperature forecasting using wavelet echo state networks with neural reconstruction. *Energy* **2013**, *57*, 382–401. [CrossRef]

30. Abdoos, A.; Hemmati, M.; Abdoos, A.A. Short term load forecasting using a hybrid intelligent method. *Knowl.-Based Syst.* **2015**, *76*, 139–147. [CrossRef]

31. Chen, J.; Li, Z.; Pan, J.; Chen, G.; Zi, Y.; Yuan, J.; Chen, B.; He, Z. Wavelet transform based on inner product in fault diagnosis of rotating machinery: A review. *Mech. Syst. Signal Process.* **2016**, *70*, 1–35. [CrossRef]

32. Zhang, Y.; Li, H.; Wang, Z.; Li, J. A preliminary study on time series forecast of fair-weather atmospheric electric field with WT-LSSVM method. *J. Electrost.* **2015**, *75*, 85–89. [CrossRef]

33. Shayeghi, H.; Ghasemi, A. Day-ahead electricity prices forecasting by a modified CGSA technique and hybrid WT in LSSVM based scheme. *Energy Convers. Manag.* **2013**, *74*, 482–491. [CrossRef]

34. Suykens, J.A.K.; Vandewalle, J. Least squares support vector machine classifiers. *Neural Process. Lett.* **1999**, *9*, 293–300. [CrossRef]

35. Sun, W.; Liang, Y. Least-squares support vector machine based on improved imperialist competitive algorithm in a short-term load forecasting model. *J. Energy Eng.* **2015**, *141*, 04014037. [CrossRef]

36. Yang, X.S.; Deb, S. Engineering optimisation by cuckoo search. *Int. J. Math. Model. Numer. Optim.* **2010**, *1*, 330–343. [CrossRef]

37. Hooshmand, R.A.; Amooshahi, H.; Parastegari, M. A hybrid intelligent algorithm based short-term load forecasting approach. *Int. J. Electr. Power Energy Syst.* **2013**, *45*, 313–324. [CrossRef]

38. Bahrami, S.; Hooshmand, R.A.; Parastegari, M. Short term electric load forecasting by wavelet transform and grey model improved by PSO (particle swarm optimization) algorithm. *Energy* **2014**, *72*, 434–442. [CrossRef]

39. Hyndman, R.J.; Koehler, A.B. Another look at measures of forecast accuracy. *Int. J. Forecast.* **2006**, *22*, 679–688. [CrossRef]

40. Niu, D.; Wang, Y.; Wu, D.D. Power load forecasting using support vector machine and ant colony optimization. *Expert Syst. Appl.* **2010**, *37*, 2531–2539. [CrossRef]

energies

MDPI

Article

Analysis and Modeling for China's Electricity Demand Forecasting Using a Hybrid Method Based on Multiple Regression and Extreme Learning Machine: A View from Carbon Emission

Yi Liang [1],*, Dongxiao Niu [1], Ye Cao [2] and Wei-Chiang Hong [3]

[1]   School of Economics and Management, North China Electric Power University, Beijing 102206, China;
      niudx@126.com
[2]   College of Management and Economy, Beijing Institute of Technology, Beijing 100081, China;
      15733221284@163.com
[3]   Department of Information Management, Oriental Institute of Technology, New Taipei 220, Taiwan;
      samuelsonhong@gmail.com
*   Correspondence: lianglouis@126.com; Tel.: +86-18810788997

Academic Editor: Javier Contreras
Received: 8 October 2016; Accepted: 4 November 2016; Published: 11 November 2016

Abstract: The power industry is the main battlefield of CO_2 emission reduction, which plays an important role in the implementation and development of the low carbon economy. The forecasting of electricity demand can provide a scientific basis for the country to formulate a power industry development strategy and further promote the sustained, healthy and rapid development of the national economy. Under the goal of low-carbon economy, medium and long term electricity demand forecasting will have very important practical significance. In this paper, a new hybrid electricity demand model framework is characterized as follows: firstly, integration of grey relation degree (GRD) with induced ordered weighted harmonic averaging operator (IOWHA) to propose a new weight determination method of hybrid forecasting model on basis of forecasting accuracy as induced variables is presented; secondly, utilization of the proposed weight determination method to construct the optimal hybrid forecasting model based on extreme learning machine (ELM) forecasting model and multiple regression (MR) model; thirdly, three scenarios in line with the level of realization of various carbon emission targets and dynamic simulation of effect of low-carbon economy on future electricity demand are discussed. The resulting findings show that, the proposed model outperformed and concentrated some monomial forecasting models, especially in boosting the overall instability dramatically. In addition, the development of a low-carbon economy will increase the demand for electricity, and have an impact on the adjustment of the electricity demand structure.

Keywords: electricity demand forecasting; multiple regression (MR); extreme learning machine (ELM); induced ordered weighted harmonic averaging operator (IOWHA); grey relation degree (GRD); carbon emission

1. Introduction

As one of the leading pioneers of national economy advancement, the electricity sector shoulders the responsibility of ensuring a stable electricity consumption and economic expansion rapidly at home and abroad [1]. Relevant to the characteristics of the electric power commodity, such as instantaneous production, transport and consumption as well as non-storability, future power demand prediction seems imperative and inevitably required. Accordingly, such sort of prediction is beneficial to the entire electricity development planning process by allowing scientifically and timely adjustment of power demand variation conditions towards sustainability [2].

With the increasing attention to climate change and greenhouse gas (GHGs) emission abatement worldwide, China has initially attempted to extend a low-carbon economy pattern, namely in the pursuit of adoption of technical progress and institutional innovation to transform energy utilization patterns, enhance energy efficiency and optimize the energy sector structure [3]. Among GHGs forms, CO_2 is on the top of list, accounting for 77% of global warming potential [4]. In China, CO_2 emissions generated by fossil energy consumption not only account for approximately 80% of total global greenhouse emissions, but also account for more than two-thirds of the responsibility for adverse greenhouse effect [5,6]. This adverse effect is representative and deteriorates seriously China' electricity sector. Regarding this, China has taken considerable countermeasures to address low-carbon issues, like climate deterioration, late environmental-protection starting of power sectors and so forth. In 2007, "Energy Saving Generation Dispatching" was published to decrease the carbon emission coefficient mainly caused by the thermal power structure [7]. Since 2013, much focus been placed on the emission-reducing effects of renewable energy sources, like zero release terms, and the National Development and Reform Commission (NDRC) in China has issued the so-called "12th five-year plan of renewable energy development" to further raise the proportion of renewable energy sources in the energy consumption mix to 15% [8] by 2020. As the largest emission-cutting participant in the clean development mechanism (CDM), China has obtained large emissions reductions from zealous participation and introduction of low-carbon technology and funds, whose checked emission reduction (CERs) reached 50% of the global share [9]. Generally, electricity demand forecasting research from the perspective of low-carbon economy proves much practical significance and practical value.

Currently, in the existing macroeconomic background, both domestic and international, numerous countries have selected appropriate variables and models to forecast electricity demand, such as Italy [10], Spain [11], USA [12], Brazil [13], Japan [14], Singapore [15], Thailand [16] and Indonesia [17]. In general, electricity demand forecasting methods can be decomposed into two aspects, namely traditional forecasting models and modern intelligent forecasting models. When it comes to traditional forecasting models, time series [18–20], regression analysis [21], Gray forecasting [22], fuzzy forecasting [23], index decomposition method [24] and so forth, are implemented widely. Pappas et al. [19,20] applied auto regressive moving average (ARMA) model to model the electricity demand loads in Greece, respectively using the Akaike corrected information criterion (AICC) and multi-model partitioning algorithm (MMPA). Hussain et al. [18] integrated Holt-winter with autoregressive integrated moving average (ARIMA) models on time series secondary data covering 1980–2011 in Pakistan, to predict overall and segmental electricity consumption. García and Carcedo [21] present an alternative analysis of electricity demand, on the basis of a simple growth rate decomposition scheme that allows vital factors behind this evolution to be identified. Similarly, Torrini et al. [22] employed the extended properties of fuzzy logic methodology to forecast the long-run electricity consumption in Brazil; while Zhao et al. [23] recommended an improved GM (1,1) model using Inner Mongolia as object. Further, multiple linear regression analysis and a quadratic regression analysis were performed deeply by Fumo et al. [24] on hourly and daily data from a research house. Inevitably, these traditional models have been comparatively proved to display a simple range of application and low-accuracy prediction thorough validated tools and simplified calculation. As for modern intelligent forecasting models, Günay [25] proposed artificial neural networks to forecast annual gross electricity demand using predicted values of social-economic indicators and climatic conditions. Son and Kim [26] applied support vector regression with particle swarm optimization algorithms to forecast the residential sector's electricity demand. Modern intelligent forecasting models have demonstrated excellent performance, including simplified regression course, transformation inference realization from training samples to predicted samples as well as avoidance towards the traditional process from induction to deduction [27]. However, they are easily trapped in over-fitting, local optima and so on. As a new type of single-hidden layer feedforward neural network primarily proposed by Huang et al. [28], extreme learning machine (ELM) embodies the features of adaptive ability, autonomic learning and optimal computation needed for unstructured and imprecise disciplines. Only by designing the suitable hidden layer nodes before training, bestowing value on input weight and partial hidden layer randomly in process, as well as simultaneously fulfilling at once without iterative, a sole optimum solution will be obtained.

Various forecasting models vary greatly from the perspective of distinct points to reflect economic variation tendencies, thus strengthening the weakness of lower accuracy using a single forecasting model. It was Bates and Granger [29] who firstly advocated combination forecasting approaches in 1969, and since then a considerable volume of studies have been conducted in many fields by domestic and overseas scholars [30–32]. The essence of combined forecasting is to solve the weighted average of single forecasting models. However, existing traditional combination forecasting models have fallen into a paralogism, namely different single forecasting models with distinguished weight coefficients, while constant combination models have unchanged weight coefficient [33]; in reality, the weight coefficient of a single forecasting model is supposed to be a function of time. Problems posed by traditional thought, are comprehensively conquered by the establishment of IOWHA operator-based forecasting models [34] in a concept of distinguished weight coefficients with the same single forecasting model over time [34,35]. Furthermore, the forecasting accuracy of the IOWHA operator shows an overdependence on the reciprocal error sum of squares which similarly is influenced by outliers to magnify the errors. Regarding this, the relevant properties of the grey relation degree (GRD) were devised and integrated with the IOWHA operator such as robust index combination, including dominance combination forecasting, non-pessimum forecasting and redundancy degree [36].

Hence, based on previous literature, a new framework of combination forecasting electricity demand model is characterized as follows: firstly, integration of GRD with the IOWHA operator to propose a new weight determination method of combination forecasting model on the basis of forecasting accuracy as induced variables; secondly, utilization of the proposed weight determination method to construct the optimal combination forecasting model based on the ELM forecasting model and multiple regression model; thirdly, three scenarios in line with the realization level of various low-carbon economy targets and dynamic simulation of the effects of low-carbon economy on future electricity demand. The remainder of this paper is organized as follows: Section 2 discusses low-carbon target scenario setting. In Section 3, a new combined GRD-IOWHA operator forecasting model is proposed. Sections 4 and 5 discuss the combination forecasting model and model results of electricity demand in China, respectively. Overall conclusions are summarized in Section 6.

2. Low-Carbon Economy Simulation Scenarios

2.1. Variation Tendency Analysis of China's Electricity Demand

By 2014, electricity consumption in China approached approximately 5626.31 million MW·h, which accounted for a quarter of world's total electricity consumption and ranked the first. Thus, electricity demand of China is representative and outperformed in terms of both applicability and feasibility.

As Figure 1 depicts (the data is sourced from the China Statistical Yearbook), in 2000–2014, the annual electricity demand of China enjoyed stable and relatively fast growth, with an average annual growth rate of 10.82%; During that period, the steepest increasing emerged in 2003, with a growth rate of 16.53%. From 2000 to 2007, electricity demand still maintained a high upward trend at an average growth rate of 13.54%; meanwhile, power demand in 2008–2009 slowed down, especially for export-oriented areas (such as East China and Guangdong at merely 5.59% and 7.21%) due to several constraint factors, including the crunch in domestic credit, Renminbi (RMB) appreciation, changes in international market demand, adjusted import-export policy, and regulatory resources, climate change, etc.

Along with the comprehensive implementation of "12th Five-Year Program", China has been accelerating the shifting in economic growth model to achieve sound and fast economic growth, together with attempts to support strategic emerging industries and upgrade traditional industries. Subsequently, the continuously adjusted consumption structure has curbed the excessive expansion of the heavy energy-consuming industry (including chemical industry, building materials, black metal smelting and smelting non-ferrous metal) and suppress China's electricity demand at a lower level. Typically in 2014, China shows a year-on-year electricity demand growth of 3.8% together with a year-on-year growth rate drop 5.12%. Under the existence of multiple uncertainties, electricity demand prediction is worthy of further exploration for prospective programming.

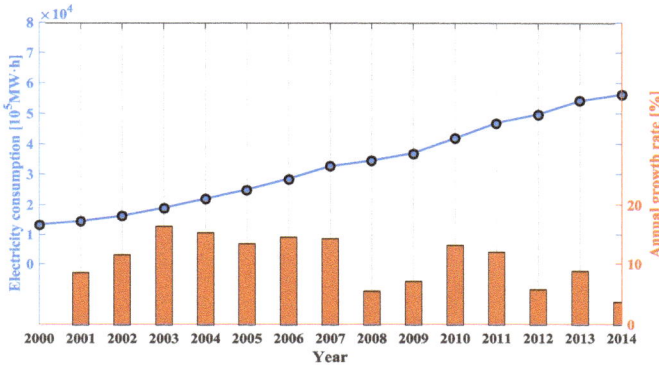

Figure 1. Annual electricity consumption of China.

2.2. Scenario Mode

With the objective to clarify the effect of energy-efficient and emission-cutting constraints on future electricity demand, three scenarios are set here to dynamically simulate future electricity demand forecasting:

(1) Baseline scenario mode. Under this mode, electricity demand growth is stimulated by economic advancement and booming population in the direction of a scheduled economic growth rate of a moderately prosperous society and population progress the same as usual.
(2) Low-carbon scenario mode. Low-carbon mode is aimed at fulfilling emission-reducing responsibilities promised during international climate talks and simultaneously promoting economic advancement by technical progress, industrial restructuring and so forth. Excluding the impact factors of economic development and population growth, electricity demand is also restrained by carbon emissions quotas.
(3) Intensified low-carbon scenario mode. Along with the thorough implementation of energy-conserved and emission-reducing policies and economic development pattern transformation towards three-low issues (low consumption, low emissions and low pollution), a low-carbon economy is well achieved by converted energy utilization patterns, enhanced energy efficiency, adjusted energy structures and so on. In this intensified mode, electricity demand is largely influenced by economic development, population growth, policy constraint and so forth.

2.3. Scenario Parameter Setting

In views of factor diversity and variability, there is a necessity to elaborate future development trend of electricity demand impact factors especially in a mid-and long term. Numerous factors are involved in China's electricity demand variation, such as economic development level, electricity price, population growth and policy constraint [24–32]. While considering the data availability and typicality, this paper merely took into account of gross domestic product (GDP), booming of the population and energy policy constraints (specifically explained in Section 2.3.3), described in Table 1.

Table 1. Historical data of scenario parameter.

Year	Electricity Demand (10^5 MW·h)	GDP (10^{12} Yuan)	Population (10^8 People)	CO$_2$ Emission Per GDP (10^4 Tons Standard Coal)	Energy Consumption Per GDP (10^4 Tons Standard Coal)
2000	13,472.4	9.98	12.67	0.97	1.49
2001	14,633.5	11.03	12.76	0.92	1.43
2002	16,331.5	12.10	12.85	0.91	1.42
2003	19,031.6	13.66	12.92	0.95	1.45
2004	21,971.4	16.07	13.00	0.94	1.44
2005	24,940.3	18.59	13.08	0.93	1.42
2006	28,588.0	21.77	13.14	0.86	1.32
2007	32,711.8	26.80	13.21	0.76	1.16
2008	34,541.4	31.68	13.28	0.65	1.01
2009	37,032.2	34.56	13.35	0.63	0.97
2010	41,934.5	40.89	13.41	0.56	0.89
2011	47,000.9	48.41	13.47	0.52	0.81
2012	49,762.6	53.41	13.54	0.48	0.75
2013	54,203.4	58.80	13.61	0.45	0.71
2014	56,263.1	63.61	13.68	0.42	0.67

2.3.1. Economic Development Level and Population

(1) GDP. Here GDP is chosen to represent economic development level. According to deepening target released in the 18th national congress of the communist party of China [37], GDP will double by 2020 with an annual growth rate at 7% roughly. Table 2 and Figure 2 illustrated GDP growth by 2020.

Table 2. Scenario parameter setting.

Year	GDP (10^{12} Yuan)	Population (10^8 People)	CO$_2$ Emission Per GDP (10^4 Tons Standard Coal)	Energy Consumption Per GDP (10^4 Tons Standard Coal)
2015	68.06	13.75	0.40	0.65
2016	72.83	13.82	0.39	0.63
2017	77.92	13.88	0.37	0.61
2018	83.38	13.95	0.36	0.59
2019	89.22	14.02	0.34	0.57
2020	95.46	14.09	0.33	0.55

Data source: NBS (National Bureau of Standards) and National Development and Reform Commission Energy Research Institute.

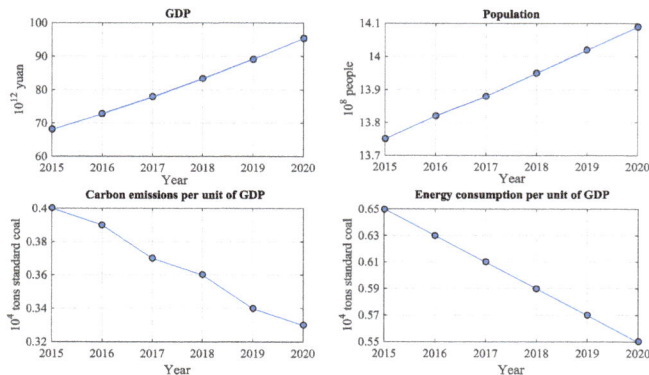

Figure 2. Scenario parameter setting.

(2) Population growth. Based on constraint conditions, population forecasting variables mainly adopted direct influential parameters, like birth rate, death rate and mobility ratio. Empirical model can be interpreted as:

$$N_t = N_{t_0} e^{k(t-t_0)} \tag{1}$$

where N_t means gross population at t; is the population base at $t = t_0$; k denotes natural population growth rate; e is the base of natural logarithms ($e = 2.718$). In line with the stable natural population growth rate over 2009–2014, we assume $k = 4.92‰$. Besides, population base is set as gross population in 2014, namely =13.68.

2.3.2. Electricity Price

Due to inexhaustive electric power system reformation and an immature electricity market, electricity price is determined by governmental macroeconomic regulation rather than an open market. Thus electricity price is failed to be predicted and ignored.

2.3.3. Energy Policy

Policy on energy conservation and emission reduction have energetically affected China's electricity demand variation and provided a more explicit target. Here CO_2 emissions per GDP and energy consumption per GDP are picked as explained variables:

$$\text{Energy consumption per GDP} = \text{Gross primary energy consumption} \div \text{GDP}$$

$$CO_2 \text{ emissions of primary energy at i} = \text{Primary energy consumption at i}$$
$$\times CO_2 \text{ emissions factor}$$

where primary energy includes coal, oil, natural gas and nuclear power, hydropower, wind power and so on; According to the National Development and Reform Commission Energy Research Institute in 2003, CO_2 emission factors of coal, oil, natural gas, nuclear power, hydropower and wind power are separately 0.7476, 0.582 5, 0.443 5, 0, 0 and 0 [38]. From the requirements of the 13th Five Year Plan [39], up to 2020, CO_2 emissions per GDP and energy consumption per GDP are reduced by 18% and 15% respectively. This study chooses the average value of 2015, namely CO_2 emissions per GDP at 3.6% and energy consumption per GDP at 3%, and then calculates their values in 2020, as Table 2 and Figure 2.

3. Combination Forecasting Model of Electricity Demand Using GRD-IOWHA Operator

3.1. Regression Analysis

3.1.1. Multiple Linear Regression

Multiple linear regression, aiming at investigating the linear relationship between dependent variable and multiple independent variables, is written as below [40]:

$$Y = \beta_0 + \beta_1 X_1 + \beta_2 X_2 + \beta_3 X_3 + \cdots + \beta_j X_j + \cdots + \beta_k X_k + \varepsilon \tag{2}$$

where k is the quantity of explanatory variable; β_j ($j = 1,2, \ldots ,k$) means regression coefficient; ε denotes the random error after eliminating the effect of independent variables on Y. Stochastic equations can be expressed as Equation (3). Besides, if X is column full rank, ordinary least squares estimate could be adopted to Equation (3), boiled down to Equation (4):

$$Y = X\beta + \varepsilon \tag{3}$$

$$\hat{\beta} = (X'X)^{-1} X'Y \tag{4}$$

3.1.2. Ridge Regression

Serious multicollinearity may lead to the failure of regression models, thus providing invalid results. Ridge regression has been exclusively used to eliminate multicollinearity by abandoning unbiasedness of least square method [41]. For the linear regression model in Equation (3), regressed parameter β can be transformed as [39]:

$$\hat{\beta}(k) = \left(X'X + kI\right)^{-1} X'Y \tag{5}$$

where $k > 0$ is ridge parameter. Varies greatly from various k, thus deeming as estimator clan. Estimator clan can be drawn by a portrait of along the k.

3.2. Extreme Learning Machine (ELM)

Different from a traditional feed forward neural network, ELM uses a non-iterative hidden layer, random selection of input weight and node and successive computed-output weight. ELM is aimed at achieving minimum training error. Excitation function G, having hidden layer N is interpreted as [28]:

$$f_{\overline{N}} = \sum_{i=1}^{\overline{N}} \beta_i G\left(a_i, b_i, x_j\right) = t_j \quad j = 1, 2, \cdots, N \tag{6}$$

where $a_i = [a_1, a_2, \cdots, a_n]^T$ is the weight vector of hidden node i; $\beta_i = [\beta_1, \beta_2, \cdots, \beta_n]^T$ means the weight vector of input node and output node; b_i is polarization of node i; denotes hidden node quantity. For simplicity, Equation (6) is transformed as:

$$H\beta = T \tag{7}$$

$$H\left(a_1, \cdots, a_{\overline{N}}, b_1, \cdots, b_{\overline{N}}, x_1, \cdots, x_N\right) = \begin{bmatrix} G\left(a_1, b_1, x_1\right) & \cdots & G\left(a_{\overline{N}}, b_{\overline{N}}, x_1\right) \\ \vdots & \ddots & \vdots \\ G\left(a_1, b_1, x_N\right) & \cdots & G\left(a_{\overline{N}}, b_{\overline{N}}, x_N\right) \end{bmatrix}_{N \cdot \overline{N}} \tag{8}$$

$$\begin{cases} \beta = \left[\beta_1^T, \cdots, \beta_{\overline{N}}^T\right]_{\overline{N} \cdot m} \\ T = \left[t_1^T, \cdots, t_N^T\right]_{N \cdot m} \end{cases} \tag{9}$$

where H means output matrix of hidden layer. Output weight can be obtained from least square solution Equation (11) of Equation (10):

$$||H\beta - T|| = ||HH^+T - T|| = \min_{\beta}||H\beta - T|| \tag{10}$$

$$\beta = H^+T \tag{11}$$

where H^+ is Moore-Penros generalized inverse matrix of .

3.3. IOWHA Operator

Supposing $\langle u_1, a_1\rangle, \langle u_2, a_2\rangle, \cdots, \langle u_n, a_n\rangle$ as two-dimensional array, $W = (w_1, w_2, \cdots, w_n)^T$ means a weighted vector related to H_w, and $\sum_{i=1}^{n} w_i = 1$. By definition, H_w points to induced ordered weighted harmonic averaging (IOWHA) operator [35] as Equation (12):

$$H_w\left(\langle u_1, a_1\rangle, \langle u_2, a_2\rangle, \cdots, \langle u_n, a_n\rangle\right) = 1 \left/ \sum_{i=1}^{n} \frac{w_i}{a_u - index\,(i)} \right. \tag{12}$$

where u_i is the induced value of a_i; $u - index(i)$ denotes the subscript of u_i. Weight coefficient w_i has nothing to do with position and size of a_i, but position of its induced value.

3.4. IOWHA Operator-Based Combination Forecasting Model

Among existence of m kinds of single forecasting models, we assume x_{it} as the forecasting value of i at t and suppose $l_1, l_2, \ldots, l_m$ as the weighted coefficient of single forecasting models in combination forecasting:

$$a_{it} = \begin{cases} 1 - |(x_t - x_{it})/x_t|, & |(x_t - x_{it})/x_t| < 1 \\ 0, & |(x_t - x_{it})/x_t| \geq 1 \end{cases} , i = 1, 2, \cdots, m, \ t = 1, 2, \cdots, N \qquad (13)$$

where a_{it} means the forecasting accuracy of model i at t; $i = 1, 2, \ldots, m$; $t = 1, 2, \ldots, N$.

Taken forecasting accuracy a_{it} as induced value of x_{it}, assumed $\langle a_{1t}, x_{1t} \rangle, \langle a_{2t}, x_{2t} \rangle, \cdots, \langle a_{mt}, x_{mt} \rangle$ as a two-dimensional array of m and arranged forecasting accuracy $a_{1t}, a_{2t}, \ldots, a_{mt}$, then Equation (14) is summarized, termed as IOWHA operator-based combination forecasting value by $a_{1t}, a_{2t}, \ldots, a_{mt}$:

$$\hat{x}_t = H(\langle a_{1t}, x_{1t} \rangle, \langle a_{2t}, x_{2t} \rangle, \cdots, \langle a_{mt}, x_{mt} \rangle) = 1 \bigg/ \sum_{i=1}^{m} \frac{l_i}{x_{a-index(it)}}, t = 1, 2, \cdots, N \qquad (14)$$

In average combination forecasting, time series is processed by selecting reciprocal error for convenience. S, reciprocal error sum squares of IOWHA operator-based combination forecasting, is written in Equation (15). Abridged weighting coefficient vector of single forecasting methods as, then we can transform Equation (15) into Equation (16) [5]:

$$S = \sum_{t=1}^{N} \left(\frac{1}{x_t} - \frac{1}{\hat{x}_t} \right)^2 = \sum_{t=1}^{N} \left(\sum_{i=1}^{m} l_i \left(\frac{1}{x_t} - \frac{1}{x_{a-index(it)}} \right) \right)^2 = \sum_{i=1}^{m} \sum_{j=1}^{m} l_i l_j \left(\sum_{t=1}^{N} e_{a-index(it)} e_{a-index(jt)} \right) \qquad (15)$$

where $e_{a-index(it)} = \frac{1}{x_t} - \frac{1}{x_{a-index(it)}}$.

$$\min S(L) = \sum_{i=1}^{m} \sum_{j=1}^{m} l_i l_j \left(\sum_{i=1}^{N} e_{a-index(it)} e_{a-index(jt)} \right)$$

$$\text{s.t.} \begin{cases} \sum_{i=1}^{m} l_i = 1 \\ l_i \geq 0, \quad i = 1, 2, \cdots, m \end{cases} \qquad (16)$$

3.5. GRD-IOWHA Operator-Based Combination Forecasting Model

IOWHA operator-based combination forecasting model, usually exploits reciprocal error sum of squares to reflect forecasting accuracy. While, reciprocal error sum of squares is easily influenced by outliers thus leading to error amplification. Regarding this, grey relation degree (GRD) is introduced to maintain robust forecasting.

Both sides of Equation (14) are handled by reciprocal like Equation (17):

$$\frac{1}{\hat{x}} = \sum_{i=1}^{m} \frac{l_i}{x_{a-index(it)}}, t = 1, 2, \cdots, N \qquad (17)$$

Seen from Equation (17):

$$\min_{1 \leq i \leq m} \frac{1}{x_{it}} = \min_{1 \leq i \leq m} \frac{1}{x_{a-index(it)}} \leq \frac{1}{\hat{x}} \leq \max_{1 \leq i \leq m} \frac{1}{x_{a-index(it)}} = \max_{1 \leq i \leq m} \frac{1}{x_{it}}, t = 1, 2, \cdots, N \qquad (18)$$

e_t is assumed as the reciprocal error between combination forecasting value and actual value at t, therefore Equation (19) appears. We can call Equation (20) GRD of reciprocal series between single forecasting method forecasting values and real values of i. Likely, Equation (21) is named the GRD of reciprocal series between IOWHA operator-based forecasting value and real values of i.

$$e_t = \frac{1}{x_t} - \frac{1}{\hat{x}_t} = \sum_{i=1}^{m} l_i \frac{1}{x_t} - \sum_{i=1}^{m} l_i \frac{1}{x_{a-index(it)}} = \sum_{i=1}^{m} l_i \left(\frac{1}{x_t} - \frac{1}{x_{a-index(it)}} \right) = \sum_{i=1}^{m} l_i e_{a-index(it)} \tag{19}$$

$$\gamma_i = \frac{1}{N} \sum_{t=1}^{N} \frac{\min\limits_{1 \le i \le m} \min\limits_{1 \le t \le N} |e_{it}| + \rho \max\limits_{1 \le i \le m} \max\limits_{1 \le t \le N} |e_{it}|}{|e_{it}| + \rho \max\limits_{1 \le i \le m} \max\limits_{1 \le t \le N} |e_{it}|} \tag{20}$$

$$\gamma = \frac{1}{N} \sum_{t=1}^{N} \frac{\min\limits_{1 \le i \le m} \min\limits_{1 \le t \le N} |e_{it}| + \rho \max\limits_{1 \le i \le m} \max\limits_{1 \le t \le N} |e_{it}|}{|e_t| + \rho \max\limits_{1 \le i \le m} \max\limits_{1 \le t \le N} |e_{it}|} \tag{21}$$

where $e_{it} = 1/x_t - 1/x_{it}$ means the reciprocal errors between forecasting values and real values of i and $\rho \in (0,1)$ is the resolution coefficient, usually at 0.5.

Based on Equation (19), GRD of reciprocal series between combination forecasting values and actual values, i.e., γ can be rewritten as below:

$$\gamma = \frac{1}{N} \sum_{t=1}^{N} \frac{\min\limits_{1 \le i \le m} \min\limits_{1 \le t \le N} |e_{it}| + \rho \max\limits_{1 \le i \le m} \max\limits_{1 \le t \le N} |e_{it}|}{\left| \sum\limits_{i=1}^{m} l_i \left(\frac{1}{x_t} - \frac{1}{x_{a-index(it)}} \right) \right| + \rho \max\limits_{1 \le i \le m} \max\limits_{1 \le t \le N} |e_{it}|} \tag{22}$$

where γ is the function of weighting coefficient vector of single forecasting model, called $\gamma(L)$. A higher γ, the more effective combination forecasting model will be. Hence, IOWHA operator-based combination forecasting model is summarized as:

$$\max \gamma(L) = \frac{1}{N} \sum_{t=1}^{N} \frac{\min\limits_{1 \le i \le m} \min\limits_{1 \le t \le N} |e_{it}| + \rho \max\limits_{1 \le i \le m} \max\limits_{1 \le t \le N} |e_{it}|}{\left| \sum\limits_{i=1}^{m} l_i \left(\frac{1}{x_t} - \frac{1}{x_{a-index(it)}} \right) \right| + \rho \max\limits_{1 \le i \le m} \max\limits_{1 \le t \le N} |e_{it}|}$$

$$\text{s.t.} \begin{cases} \sum\limits_{i=1}^{m} l_i = 1 \\ l_i \ge 0, \quad i = 1, 2, \cdots, m \end{cases} \tag{23}$$

Plugging into Equation (24) to perform GRD-IOWHA operator-based combination forecasting:

$$\hat{x}_t = H \left(\langle a_{1t}, x_{1t} \rangle, \langle a_{2t}, x_{2t} \rangle, \cdots, \langle a_{mt}, x_{mt} \rangle \right) = 1 / \sum_{i=1}^{m} \frac{l_i^*}{x_{a-index(it)}}, \quad t = N+1, N+2, \cdots, \tag{24}$$

where during interval [N+1,N+2, ... ,], the size of forecasting accuracy series $a_{1t}, a_{2t}, \ldots, a_{mt}$, is determined by the distance to average fitting accuracy. In other words, the forecasting accuracy in interval $N + k$ is substituted by average fitting accuracy $1/k \cdot \sum_{t=N-k+1}^{N} a_{it}$ of step k.

Regression analysis is termed as RA, while grey relation degree and modified IOWHA operator is short for GRD-IOWHA operator. Thus far, based on modified GRD-IOWHA operator, combination forecasting modeling is constituted by multiple regression as well as ELM, and completely fulfilled. Figure 3 depicts the operational process concretely, where the left demonstrates two single forecasting modeling and the right explains combination forecasting modeling.

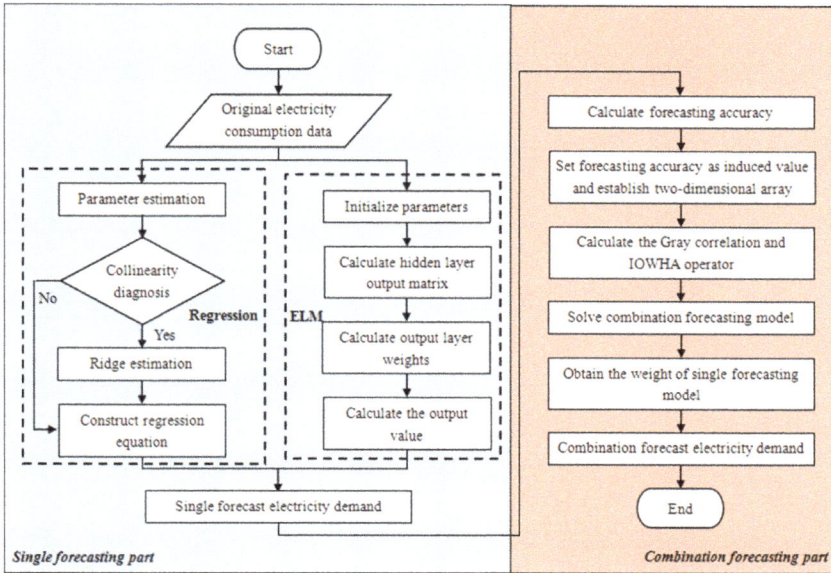

Figure 3. GRD-IOWHA operator-based combination forecasting modeling process.

4. Electricity Demand Forecasting in China

This section took full advantage of the above-proposed combination forecasting model to predict China's electricity demand under three types of low-carbon scenarios. Among that, for MR forecasting, we use data in 2000–2008 to simulate and data in 2009–2014 to test. The same occurs for ELM, where training sample is derived from data in 2000–2008, and test sample is from 2009–2014.

4.1. Baseline Scenario Forecasting

4.1.1. Forecasting of RA and ELM

Let GDP be x_1 and population be x_2, linear regression model is boiled down to the following:

$$y = 427.31x_1 + 21559.93x_2 - 264862.86$$

Deduced from calculation, modified fitting degree $R^2 = 0.997$. Moreover, in the significance level of $\alpha = 0.05$, statistics $F = 2406.802 > F_{0.05}(2,12) = 2.81$ and each statistics $T > t_{0.025}(12) = 2.179$, respectively, which means that regressed model has passed significance testing and embodies a better imitative effect. Tables 3 and 4 separately discussed test process using data from 2009–2014 and single forecasting covering data from 2015–2020.

Referencing the achievements of Huang et al. [28], this paper adopted the Matlab software to compile the ELM toolkit, together with the "Sigmoid function" to activate neurons in the hidden layer. The number of neurons of the hidden layer is set at 9. As for ELM forecasting in Table 3, data covering 2000–2008 is plugged in for training and data in 2009–2014 to test the well-trained model; Then, single forecasting of China's electricity demand in 2015–2020 is carried out, as shown in Table 4.

Table 3. Single test result in baseline scenario.

Year	Electricity Demand (10^5 MW·h)	RA		ELM	
		Value (10^5 MW·h)	Forecasting Accuracy	Value (10^5 MW·h)	Forecasting Accuracy
2009	37,032.2	37,730.04	0.9812	36,680.39	0.9905
2010	41,934.5	41,728.51	0.9951	42,509.00	0.9863
2011	47,000.9	46,235.47	0.9837	47,442.71	0.9906
2012	49,762.6	49,881.22	0.9976	48,473.75	0.9741
2013	54,203.4	53,693.62	0.9906	53,276.52	0.9829
2014	56,263.1	57,258.17	0.9823	55,908.64	0.9937

Table 4. Single forecasting result in baseline scenario.

Year	RA Forecasting (10^5 MW·h)	ELM Forecasting (10^5 MW·h)
2015	60,670.01	58,713.27
2016	64,215.07	63,592.23
2017	67,687.05	66,181.66
2018	71,527.12	70,297.18
2019	75,530.34	74,664.38
2020	79,708.15	78,938.27

4.1.2. GRD-IOWHA Forecasting

Two single forecasting results are exploited to construct forecasting accuracy and relevant in sample interval, $t = 1,2, \ldots ,6$. The IOWHA operator-based forecasting value is displayed as below:

$$\hat{x}_1 = H(\langle a_{11}, x_{11} \rangle, \langle a_{21}, x_{21} \rangle) = 1/(l_1/36680.39 + l_2/37730.04)$$

$$\hat{x}_2 = H(\langle a_{12}, x_{12} \rangle, \langle a_{22}, x_{22} \rangle) = 1/(l_1/41728.51 + l_2/42509.00)$$

$$\hat{x}_3 = H(\langle a_{13}, x_{13} \rangle, \langle a_{23}, x_{23} \rangle) = 1/(l_1/47442.71 + l_2/46235.47)$$

$$\hat{x}_4 = H(\langle a_{14}, x_{14} \rangle, \langle a_{24}, x_{24} \rangle) = 1/(l_1/49881.22 + l_2/48473.75)$$

$$\hat{x}_5 = H(\langle a_{15}, x_{15} \rangle, \langle a_{25}, x_{25} \rangle) = 1/(l_1/53693.62 + l_2/53276.52)$$

$$\hat{x}_6 = H(\langle a_{16}, x_{16} \rangle, \langle a_{26}, x_{26} \rangle) = 1/(l_1/55908.64 + l_2/57258.17)$$

where l_1 and l_2 show weighting coefficients of two single forecasting models in combination forecasting.

With its direct substitution into Equation (23), the most effective weight coefficient of combination forecasting model is expressed as below with $\rho = 0.5$.

$$l_1^* = 0.7325, \quad l_2^* = 0.2675$$

Taking the average accuracy of former 6 as each single forecasting accuracy, we can obtain the combination forecasting results of China's electricity demand in baseline scenario covering 2015–2020, shown in Table 5.

Table 5. Combination forecasting results in baseline scenario of 2015–2020.

Year	GRD-IOWHA Forecasting (10^5 MW·h)
2015	60,133.92
2016	64,047.27
2017	67,277.69
2018	71,193.91
2019	75,296.73
2020	79,500.74

4.2. Low-Carbon Scenario Forecasting

4.2.1. Forecasting of RA and ELM

In order to eliminate the multicollinearity of selected variables, the low-carbon scenario and intensified scenario necessarily employed ridge regression to achieve efficient fitting. Similarly, let GDP be x_1, population be x_2 and CO_2 emissions per GDP be x_3. Besides, take the logarithm term to remove variable heteroscedasticity. SPSS 20.0 software was used to conduct the ridge regression shown in Figures 4 and 5.

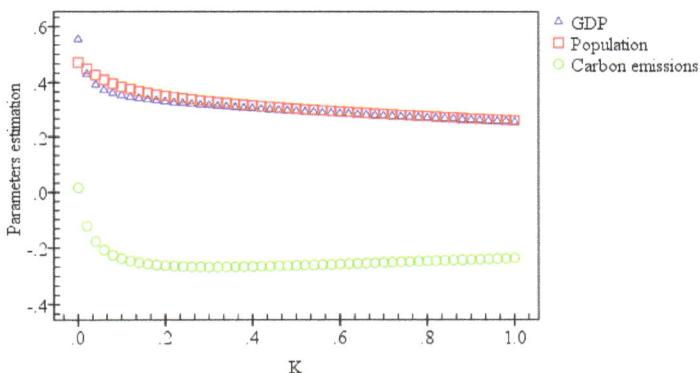

Figure 4. Ridge trace of variables.

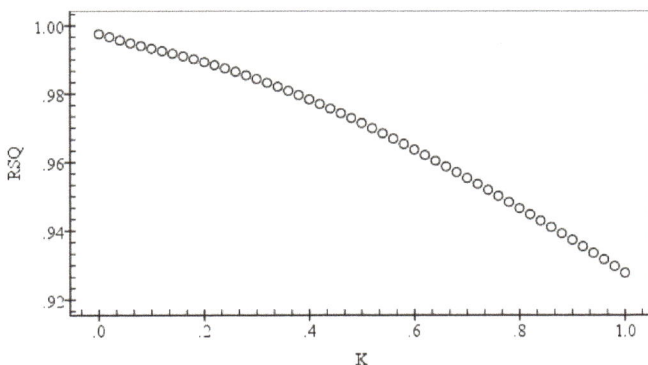

Figure 5. Determination coefficient and *K* value.

From the ridge trace, when K is nearly close to 0.2, all parameters tend to be stable; Even when K exceeds 0.2, the determination coefficient presents a stable decline without drastic fluctuation. Setting $K = 0.2$ and $R^2 = 0.9709$, the ridge regression model is fitted as follows:

$$\ln y = 0.28\ln x_1 + 7.86\ln x_2 - 0.27\ln x_3 - 10.98$$

In the significance level of $\alpha = 0.05$, statistics $F = 156.913 > F_{0.05}(3,12) = 2.61$ and each statistics $T > t_{0.05}(12) = 2.179$ all demonstrated that regressed model had passed through significance testing with a well-fitting level, as shown in Tables 6 and 7. Results of ELM method forecasting are shown there also.

Table 6. Single test result in low-carbon scenario.

Year	Electricity Demand (10^5 MW·h)	RA		ELM	
		Value (10^5 MW·h)	Forecasting Accuracy	Value (10^5 MW·h)	Forecasting Accuracy
2009	37,032.2	36,540.79	0.9867	37,683.97	0.9824
2010	41,934.5	40,878.39	0.9748	42,408.36	0.9887
2011	47,000.9	45,390.34	0.9657	47,461.51	0.9902
2012	49,762.6	49,709.48	0.9989	50,265.20	0.9899
2013	54,203.4	53,914.26	0.9947	55,531.38	0.9755
2014	56,263.1	58,759.43	0.9556	56,893.25	0.9888

Table 7. Single forecasting result in low-carbon scenario.

Year	RA Forecasting (10^5 MW·h)	ELM Forecasting (10^5 MW·h)
2015	62,954.49	59,313.86
2016	67,435.32	64,549.11
2017	71,839.90	68,457.73
2018	76,984.74	72,365.06
2019	82,518.03	79,509.00
2020	88,473.96	85,959.43

4.2.2. GRD-IOWHA Operator-Based Combination Forecasting

Iterative steps like above-mentioned, GRD-IOWHA operator-based combination forecasting is summarized as below:

$$\hat{x}_1 = H(\langle a_{11}, x_{11} \rangle, \langle a_{21}, x_{21} \rangle) = 1/(l_1/36540.79 + l_2/37683.97)$$

$$\hat{x}_2 = H(\langle a_{12}, x_{12} \rangle, \langle a_{22}, x_{22} \rangle) = 1/(l_1/42408.36 + l_2/40878.39)$$

$$\hat{x}_3 = H(\langle a_{13}, x_{13} \rangle, \langle a_{23}, x_{23} \rangle) = 1/(l_1/47461.51 + l_2/45390.34)$$

$$\hat{x}_4 = H(\langle a_{14}, x_{14} \rangle, \langle a_{24}, x_{24} \rangle) = 1/(l_1/49709.48 + l_2/50265.20)$$

$$\hat{x}_5 = H(\langle a_{15}, x_{15} \rangle, \langle a_{25}, x_{25} \rangle) = 1/(l_1/53914.26 + l_2/55531.38)$$

$$\hat{x}_6 = H(\langle a_{16}, x_{16} \rangle, \langle a_{26}, x_{26} \rangle) = 1/(l_1/56893.25 + l_2/58759.43)$$

With utilization of the optimal tool in the Matlab software, the combination forecasting model shows the most powerful coefficient, are shown as below. Future electricity demand in China is predicted in Table 8.

$$l_1^* = 0.6981, \quad l_2^* = 0.3019$$

Table 8. Combination forecasting result in low-carbon scenario.

Year	GRD-IOWHA Forecasting (10^5 MW·h)
2015	60,367.81
2016	65,394.08
2017	69,444.76
2018	73,700.24
2019	80,394.04
2020	86,703.37

4.3. Forecasting in Reinforced Low-Carbon Scenario

4.3.1. Forecasting of RA and ELM

Likewise, let GDP be x_1, population be x_2 and CO_2 emissions per GDP be x_3, energy consumption per GDP be x_4. Taking variables in logarithm terms, the ridge trace and K variation are displayed in Figures 6 and 7.

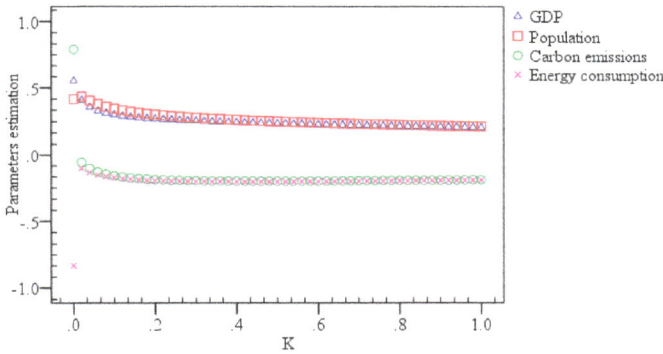

Figure 6. Ridge trace of variables.

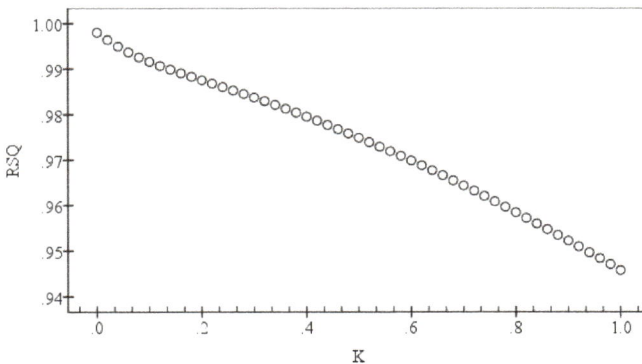

Figure 7. Determined coefficient and K value.

Based on ridge trace, when K is nearly close to 0.2, all parameters tend to be stable; Even when K exceeds 0.2, the determined coefficient presents a stable declination without drastic fluctuation. Setting $K = 0.2$ and $R^2 = 0.9623$, the ridge regression model is fitted as follows:

$$\ln y = 0.25\ln x_1 + 7.29\ln x_2 - 0.18\ln x_3 - 0.22\ln x_4 - 9.36$$

In the significance level of $\alpha = 0.05$, statistics $F = 90.430 > F_{0.05}(3,12) = 2.61$ and each statistics $T > t_{0.05}(12) = 2.179$ all demonstrated that the regressed model had passed through significance testing with a well-fitting level, as shown in Tables 9 and 10. Also, like the previous scenario's parameter setting, test results and forecasting results of ELM method are shown here.

Table 9. Single test result in reinforced low-carbon scenario.

Year	Electricity Demand (10^5 MW·h)	RA		ELM	
		Value (10^5 MW·h)	Forecasting Accuracy	Value (10^5 MW·h)	Forecasting Accuracy
2009	37,032.2	36,565.66	0.9874	36,295.26	0.9801
2010	41,934.5	41,044.15	0.9788	42,639.00	0.9832
2011	47,000.9	45,822.55	0.9749	47,673.01	0.9857
2012	49,762.6	50,249.02	0.9902	50,270.18	0.9898
2013	54,203.4	54,581.74	0.9930	54,837.58	0.9883
2014	56,263.1	59,514.14	0.9422	57,697.81	0.9745

Table 10. Single forecasting result in reinforced low-carbon scenario.

Year	RA forecasting (10^5 MW·h)	ELM forecasting (10^5 MW·h)
2015	63,665.11	64,261.28
2016	68,092.73	68,235.02
2017	72,466.81	73,265.20
2018	77,553.93	77,386.74
2019	83,027.15	82,835.97
2020	88,921.55	88,110.93

4.3.2. GRD-IOWHA Operator-Based Combination Forecasting

Similar forecasting process to above-mentioned section, IOWHA operator-based combination forecasting results display as below:

$$\hat{x}_1 = H(\langle a_{11}, x_{11} \rangle, \langle a_{21}, x_{21} \rangle) = 1/(l_1/36565.66 + l_2/36295.26)$$

$$\hat{x}_2 = H(\langle a_{12}, x_{12} \rangle, \langle a_{22}, x_{22} \rangle) = 1/(l_1/42639.00 + l_2/41044.15)$$

$$\hat{x}_3 = H(\langle a_{13}, x_{13} \rangle, \langle a_{23}, x_{23} \rangle) = 1/(l_1/47673.01 + l_2/45822.55)$$

$$\hat{x}_4 = H(\langle a_{14}, x_{14} \rangle, \langle a_{24}, x_{24} \rangle) = 1/(l_1/50249.02 + l_2/50270.18)$$

$$\hat{x}_5 = H(\langle a_{15}, x_{15} \rangle, \langle a_{25}, x_{25} \rangle) = 1/(l_1/54581.74 + l_2/54837.58)$$

$$\hat{x}_6 = H(\langle a_{16}, x_{16} \rangle, \langle a_{26}, x_{26} \rangle) = 1/(l_1/57697.81 + l_2/59514.14)$$

With utilization of optimal took it in Matlab software, combination forecasting model shows the most powerful coefficient, shown as below. Future electricity demand in China is predicted in Table 11.

$$l_1^* = 0.6459, \quad l_2^* = 0.3541$$

Table 11. Combination forecasting result in low-carbon scenario.

Year	GRD-IOWHA Forecasting (10^5 MW·h)
2015	64,048.9
2016	68,184.57
2017	72,980.49
2018	77,445.86
2019	82,903.57
2020	88,396.27

5. Results and Discussions

Deduced from China's electricity demand forecasting results under various scenarios, further discussion is concluded from four perspectives.

(1) GRD-IOWHA operator-based combination forecasting model outperformed each single forecasting model notably. Figure 8 demonstrates the forecasting accuracy comparison of single forecasting models covering testing data in 2009–2014, where Scenario 1 means baseline scenario, Scenario 2 represents the low-carbon scenario and Scenario 3 in the intensified low-carbon scenario. Single forecasting models provide various forecasting accuracy at various moments. More specifically, in the baseline scenario, the ELM model shows a superior forecasting accuracy of electricity demand than the RA model in 2009, 2011 and 2014; while the RA model is much better in 2010, 2012 and 2013. In the low-carbon scenario, the RA model provides better forecasting accuracy than the ELM model, namely 2009, 2012 and 2013; while the ELM model predicts electricity demand overwhelmingly in other years (2010, 2011 and 2014). In the intensified low-carbon scenario, the RA model provides higher forecasting accuracy in 2009, 2012 and 2013 and lower forecasting accuracy in 2010, 2011 and 2014. Generally, the proposed GRD-IOWHA operator-based combination forecasting model concentrates the advantages of various single forecasting models, namely higher weight coefficient in higher single forecasting accuracy and vice versa. According to Equations (20) and (21), Table 12 represents grey relation degree comparison, from 2009 to 2014, in various scenarios between single forecasting model and GRD-IOWHA operator-based combination forecasting model. Findings show that grey relation degree of three scenarios in GRD-IOWHA operator-based combination forecasting model is better than that of single forecasting models. Thus, the proposed combination model belongs to the dominated forecasting combination model [36].

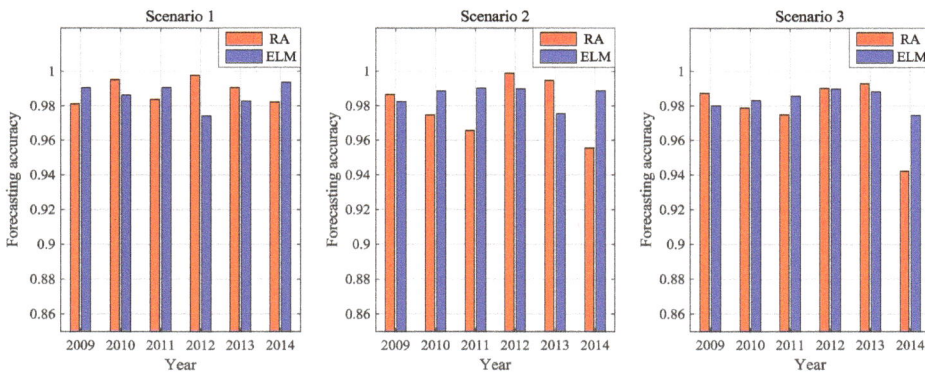

Figure 8. Forecasting accuracy comparison of single forecasting models in various scenario.

Table 12. Grey relation degree comparison in various scenarios.

	Scene 1		
Model	**RA**	**ELM**	**GRD-IOWHA**
Grey correlation value	0.6661	0.5945	0.9693
	Scene 2		
Model	**RA**	**ELM**	**GRD-IOWHA**
Grey correlation value	0.5804	0.6078	0.8532
	Scene 3		
Model	**RA**	**ELM**	**GRD-IOWHA**
Grey correlation value	0.7113	0.7503	0.9077

(2) The proposed GRD-IOWHA operator-based combination forecasting model predicts accurately and more truly than the basic IOWHA operator-based combination forecasting model [35] and the traditional combination forecasting (TCF) model [33], namely each single forecasting model with unchanged weight coefficient. In order to compare typical combination forecasting models effectively, this section compares the measured IOWHA operator-based combination forecasting model and traditional combination models shown in Table 13.

Table 13. Comparison of model forecasting result.

Year	Scene 1		Scene 2		Scene 3	
	IOWHA	**TCF**	**IOWHA**	**TCF**	**IOWHA**	**TCF**
2009	37,197.81	37,268.19	37,103.58	37,386.74	36,429.97	36,297.96
2010	42,115.14	42,071.93	41,629.32	42,010.57	41,826.38	42,623.05
2011	46,831.31	46,766.66	46,402.83	46,923.01	46,729.47	47,654.51
2012	49,167.41	49,261.93	49,985.8	50,120.71	50,259.6	50,269.97
2013	53,484.26	53,510.1	54,710.87	55,110.93	54,709.36	54,835.02
2014	56,575.36	56,664.38	57,811.28	57,378.46	58,591.9	57,715.97

Unit: 10^5 MW·h.

According to the evaluating principle of forecasting effect, the following dimensions are selected as the evaluation index system, including *RE, SSE, MSE, MAE, MAPE, MSPE*. Concretely, only *RE* is used to reflect single forecasting model effect. Figure 9 illustrates electricity demand forecasting in 2009–2014 using GRD-IOWHA operator-based combination forecasting model, IOWHA operator-based combination forecasting model and traditional combination forecasting model separately:

$$\text{Relative error}: \ RE = (\hat{x}_t - x_t)/x_t$$

$$\text{Error of sum square}: \ SSE = \sum_{t=1}^{N}(x_t - \hat{x}_t)^2$$

$$\text{Mean square error}: \ MSE = \frac{1}{N}\sqrt{\sum_{t=1}^{N}(x_t - \hat{x}_t)^2}$$

$$\text{Mean absolute error}: \ MAE = \frac{1}{N}\sum_{t=1}^{N}|x_t - \hat{x}_t|$$

$$\text{Mean Absolute Percentage Error}: \ MAPE = \frac{1}{N}\sum_{t=1}^{N}|(x_t - \hat{x}_t)/x_t|$$

where x_t denotes the actual demand value, presents the predicted value.

Compared with the other forecasting models, the proposed GRD-IOWHA operator-based combination forecasting model is rather close to actual values. From Figure 10, under the distinguished scenario, the relative error value of the proposed GRD-IOWHA operator-based combination forecasting model is in much lower interval and fluctuates slightly, followed by IOWHA operator-based combination forecasting model or traditional combination forecasting model, worst in two single forecasting model. In a word, proposed GRD-IOWHA operator-based combination forecasting model perform more superiority in decreasing forecasting error fluctuation and risk of tech-economic decision making. Furthermore, Figure 11 demonstrates the overall forecasting evaluation result of various forecasting model, especially being satisfactory and optimal condition in index *SSE*, *MSE*, *MAE* and *MAPE*. Yet exceptional situations still exist, like lower *SSE* and *MSE* in a traditional forecasting model than the GRD-IOWHA operator-based combination forecasting model under intensified low-carbon scenario due to larger forecasting error caused by single forecasting models. In spite of this, the proposed combination forecasting model outperformed both in effectiveness and feasibility as a whole.

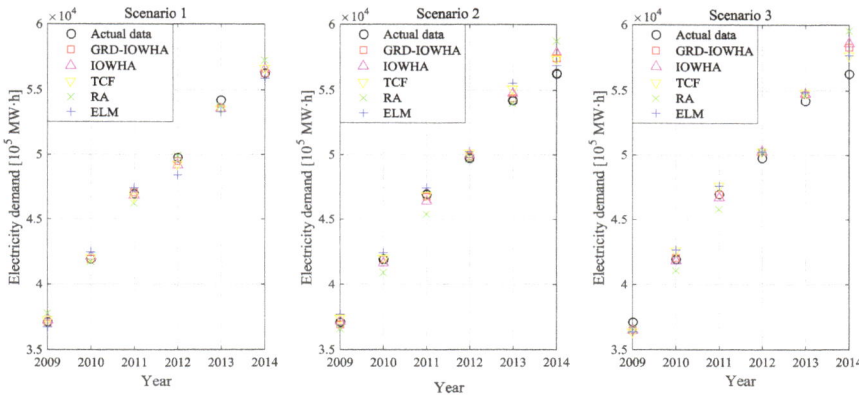

Figure 9. Electricity demand forecasting of various model in 2009–2014.

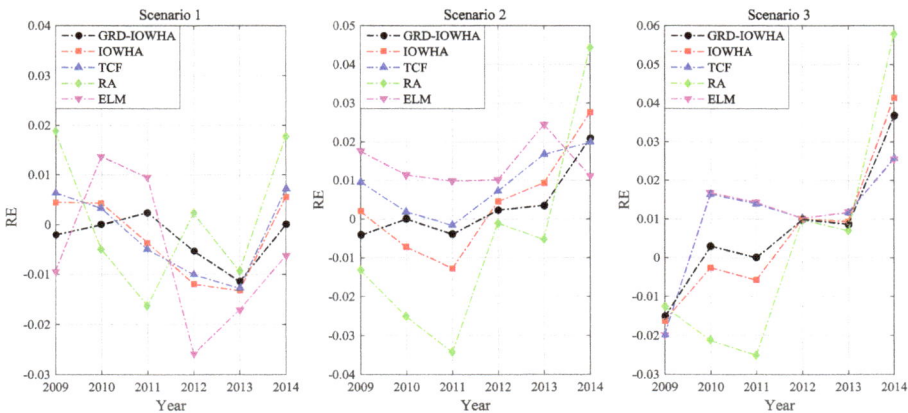

Figure 10. Relative error forecasting of various model in 2009–2014.

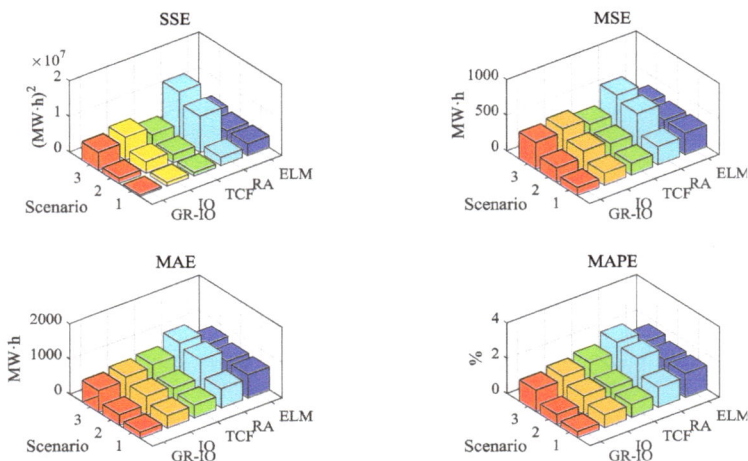

Figure 11. Evaluation index of various forecasting model.

(3) Low-carbon economy advancement contributes to augment electricity demand in China. Known from Figure 12, electricity demand under energy restriction, i.e., low-carbon scenario and intensified low-carbon scenario, is greater than that under unrestricted energy use. Hence, China will strive to cut down the utilization of high-emission releasing resources, like coal, oil, natural gas and so on as well as explore the substituent effect of electricity. Under energy restriction circumstances, electricity demand in an unchanged energy efficiency scenario is higher than that of continually improved energy efficiency, thus emission-cutting emphasis lies in energy structure optimization and electricity demand increasing. However, if China initially promises a lower energy efficiency (energy consumption per GDP), like 15% declining by 2020 rather in 2015, pressure on China's electricity demand will be cut down tremendously.

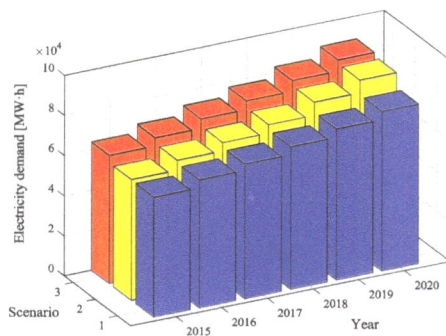

Figure 12. Electricity demand forecasting trend of various model in 2015–2020.

(4) A low-carbon economy causes a structural variation of electricity demand. Increasing electricity demand is mainly involved in renewable clean energy, like water power, nuclear power and wind power. With respect to a low-carbon scenario, i.e., unchanged energy efficiency, incremental 72.0263 million MW·h electricity demand is also chiefly centralized in renewable clean energy electricity demand compared with the baseline scenario. Despite the continuous effort on electricity structure adjustment and decreasing the ratio of coal power, the coal power ratio will not fall sharply for a

time behind the reason of over-dependence on electricity and abundant coal resources. Illustrated in baseline scenario of Figure 13, power generation is presumed to be 70% coal power ratio and 30% in water power, nuclear power and wind power, which accounts for 238.5022 million MW·h in 2020. Due to the constrained energy policy, under the unchanged energy efficiency situation, electricity demand from clean energy approaches nearly 2601.1011 million MW·h by 2020, which accounts for 32.72% of total electricity demand, while the coal power ratio decreased to 67.28%. Therefore, the low-carbon economy has affected both electricity demand and its structure variation.

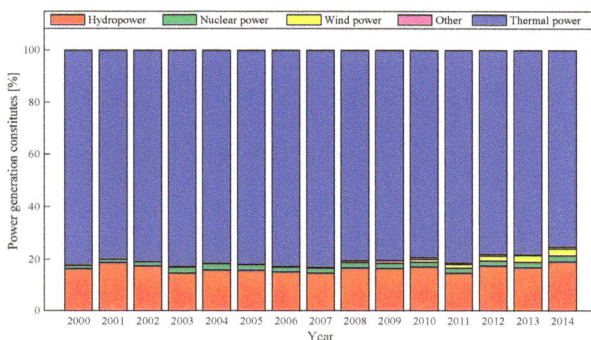

Figure 13. Constitution and proportion of China's annual power generation.

6. Conclusions

In this study, a new framework of combination forecasting electricity demand model is characterized as follows: (1) Integration of a grey relation degree with an induced ordered weighted harmonic averaging operator to propose a new weight determination method of combination forecasting model on basis of forecasting accuracy as induced variables; (2) utilization of the proposed weight determination method to construct the optimal combination forecasting model based on an extreme learning machine forecasting model and multiple regression model; (3) three scenarios in line with realization level of various low-carbon economy targets and dynamic simulation of the effects of a low-carbon economy on future electricity demand.

Resultant findings are obtained and clarified in detail: (1) the grey relation degree of reciprocal series between proposed combination forecasting value and actual values is better than the single forecasting models studied in this paper and corresponds to an optimal combination forecasting model; (2) the proposed combination forecasting model outperformed and concentrated the advantages of some monomial forecasting models, especially in boosting the overall instability dramatically and providing reliable decision basis; (3) the energetic progress of a low-carbon economy causes an increase in electricity demand and the relevant structure adjustment of electricity demand, especially in the increasing demand of clean energy. Above all, this study is aimed at providing a reference for future power planning issues in China.

Acknowledgments: This work is supported by the Natural Science Foundation of China (Project No. 71471059).

Author Contributions: Yi Liang designed this research and wrote this paper; Dongxiao Niu and Wei-Chiang Hong provided professional guidance; and Ye Cao collected all the data and translated this paper.

Conflicts of Interest: The authors declare no conflict of interest.

References

1. Sevik, S. An analysis of the current and future use of natural gas-fired power plants in meeting electricity energy needs: The case of Turkey. *Renew. Sustain. Rev.* **2015**, *52*, 572–586. [CrossRef]

2. O'Connell, N.; Pinson, P.; Madsen, H. Benefits and challenges of electrical demand response: A critical review. *Renew. Sustain. Rev.* **2014**, *39*, 686–699. [CrossRef]
3. Fan, D.; Wang, S.; Zhang, W. Research on Prediction of China's Electric Power Demand under Low-Carbon Economy Target. *Power Syst. Technol.* **2012**, *36*, 19–25. (In Chinese)
4. Kang, C.; Chen, Q.; Xia, Q. Prospects of Low-Carbon Electricity. *Power Syst. Technol.* **2009**, *33*, 1–7. (In Chinese)
5. Zeng, M.; Yang, Y.; Wang, L.; Sun, J. The power industry reform in China 2015: Policies, evaluations and solutions. *Renew. Sustain. Energy Rev.* **2016**, *57*, 94–110. [CrossRef]
6. Yang, L.; Lin, B. Carbon dioxide-emission in China's power industry: Evidence and policy implications. *Renew. Sustain. Energy Rev.* **2016**, *60*, 258–267. [CrossRef]
7. Yun, L.; Ping, H. The Policy Interpretation of the Energy Conservation Power Generating and Distribution Code. *J. Electr. Power* **2008**, *23*, 215–217. (In Chinese)
8. National Development and Reform Commission. *The Twelfth Five Years Plan of Renewable Energy Development*; National Development and Reform Commission: Beijing, China, 2013. (In Chinese)
9. Zhang, C.; Yan, J. CDM's influence on technology transfers: A study of the implemented clean development mechanism projects in China. *Appl. Energy* **2015**, *158*, 355–365. [CrossRef]
10. D'Errico, M.C.; Bollino, C.A. Bayesian Analysis of Demand Elasticity in the Italian Electricity Market. *Sustainability* **2015**, *7*, 12127–12148. [CrossRef]
11. Hernandez, L.; Baladron, C.; Aguiar, J.M. Classification and Clustering of Electricity Demand Patterns in Industrial Parks. *Energies* **2012**, *5*, 5215–5228. [CrossRef]
12. Schweizer, V.J.; Morgan, M.G. Bounding US electricity demand in 2050. *Technol. Forecast. Soc. Chang.* **2016**, *105*, 215–223. [CrossRef]
13. Trotter, I.M.; Bolkesjø, T.F.; Féres, J.G.; Hollanda, L. Climate Change and Electricity Demand in Brazil: A Stochastic Approach. *Energy* **2016**, *102*, 596–604. [CrossRef]
14. Kishita, Y.; Yamaguchi, Y.; Umeda, Y. Describing Long-Term Electricity Demand Scenarios in the Telecommunications Industry: A Case Study of Japan. *Sustainability* **2016**, *8*, 52. [CrossRef]
15. Seung, J.O.; Kim, C.N.; Kyaw, T.; Wongee, C.; Kian, J.E.C. Forecasting Long-term Electricity Demand for Cooling of Singapore's Buildings Incorporating an Innovative Air-conditioning Technology. *Energy Build.* **2016**, in press.
16. Kandananond, K. Forecasting Electricity Demand in Thailand with an Artificial Neural Network Approach. *Energies* **2011**, *4*, 1246–1257. [CrossRef]
17. Suhono, S. Long-term Electricity Demand Forecasting of Sumatera System Based on Electricity Consumption Intensity and Indonesia Population Projection 2010–2035. *Energy Procedia* **2015**, *68*, 455–462. [CrossRef]
18. Pappas, S.S.; Ekonomou, L.; Moussas, V.C.; Karampelas, P.; Katsikas, S.K. Adaptive load forecasting of the Hellenic electric grid. *J. Zhejiang Univ. Sci. A* **2008**, *9*, 1724–1730. [CrossRef]
19. Pappas, S.S.; Ekonomou, L.; Karamousantas, D.C.; Chatzarakis, G.E.; Katsikas, S.K.; Liatsis, P. Electricity demand loads modeling using autoregressive moving average (ARMA) models. *Energy* **2008**, *33*, 1353–1360. [CrossRef]
20. Hussain, A.; Rahman, M.; Memon, J.A. Forecasting electricity consumption in Pakistan: The way forward. *Energy Policy* **2016**, *90*, 73–80. [CrossRef]
21. Pérez-García, J.; Moral-Carcedo, J. Analysis and long term forecasting of electricity demand trough a decomposition model: A case study for Spain. *Energy* **2016**, *97*, 127–143. [CrossRef]
22. Torrini, F.C.; Souza, R.C.; Cyrino-Oliveira, F.L. Long term electricity consumption forecast in Brazil: A fuzzy logic approach. *Socio-Econ. Plan. Sci.* **2016**, *70*, 101–114. [CrossRef]
23. Zhao, H.R.; Zhao, H.R.; Guo, S. Using GM (1,1) Optimized by MFO with Rolling Mechanism to Forecast the Electricity Consumption of Inner Mongolia. *Appl. Sci.* **2016**, *6*, 20. [CrossRef]
24. Fumo, N.; Biswas, M.A.R. Regression analysis for prediction of residential energy consumption. *Renew. Sustain. Rev.* **2015**, *47*, 332–343. [CrossRef]
25. Günay, M.E. Forecasting annual gross electricity demand by artificial neural networks using predicted values of socio-economic indicators and climatic conditions: Case of Turkey. *Energy Policy* **2016**, *90*, 92–101. [CrossRef]
26. Son, H.; Kim, C. Short-term forecasting of electricity demand for the residential sector using weather and social variables. *Resour. Conserv. Recycl.* **2016**, in press. [CrossRef]

27. Ekonomou, L.; Oikonomou, D.S. Application and comparison of several artificial neural networks for forecasting the Hellenic daily electricity demand load. In Proceedings of the 7th WSEAS International Conference on Artificial Intelligence, Knowledge Engineering and Data Bases (AIKED'08), Cambridge, UK, 20–22 February 2008; pp. 67–71.

28. Huang, G.B.; Zhu, Q.Y.; Siew, C.K. Extreme learning machine: A new learning scheme of feedforward neural networks. In Proceedings of the 2004 IEEE International Joint Conference on Neural Networks, Budapest, Hungary, 25–29 July 2004; pp. 985–990.

29. Bates, J.M.; Granger, C.W.J. Combination of forecasts. *Oper. Res. Q.* **1969**, *20*, 451–468. [CrossRef]

30. Mao, L.; Yao, J.; Jin, Y.; Chen, H.; Li, W.; Guan, S. Theoretical Study of Combination Model for Medium and Long Term Load Forecasting. *Proc. CSEE* **2010**, *30*, 53–59. (In Chinese)

31. Chen, H.; Sheng, Z. A Kind of New Combination Forecasting Method Based on Induced Ordered Weighted Geometric Averaging (IOWGA) Operator. *J. Ind. Eng. Eng. Manag.* **2005**, *19*, 36–39. (In Chinese)

32. Zhao, W.; Wang, J.; Lu, H. Combining forecasts of electricity consumption in China with time-varying weights updated by a high-order Markov chain model. *Omega* **2014**, *45*, 80–91. [CrossRef]

33. Chen, H. *The Effectiveness Theory and Its Application of Combination Forecasting*; Science Press Ltd.: Beijing, China, 2008.

34. Chen, H.; Liu, C. A Kind of Combination Forecasting Method Baesd on Induced Ordered Weighted Averaging (IOWA) Operators. *Forecasting* **2003**, *22*, 61–65. (In Chinese)

35. Chen, H.; Liu, C.; Sheng, Z. Induced Ordered Weighted Harmonic Averaging (IOWHA) Operator and Its Application to Combination Forecasting Method. *Chin. J. Manag. Sci.* **2004**, *12*, 35–40. (In Chinese)

36. Chen, H.; Zhao, J.; Liu, C. Properties of combination forecasting model based on degree of grey incidence. *J. Southeast Univ. (Natl. Sci. Ed.)* **2004**, *34*, 130–134. (In Chinese)

37. Hu, J. Firmly March on the Path of Socialism with Chinese Characteristics and Strive to Complete the Building of a Moderately Prosperous Society in All Respects. In Proceedings of the Report to the 18th National Congress of the Communist Party of China, Beijing, China, 8 November 2012.

38. National Bureau of Statistics of the People's Republic of China. *China Statistical Yearbook 2015*; China Statistics Press: Beijing, China, 2016.

39. Xinhua News Agency. *The Thirteenth Five-Year Plan for National Economic and Social Development of the People's Republic of China*; Xinhua News Agency: Beijing, China, 2016.

40. Laicane, I.; Blumberga, D.; Blumberga, A.; Rosa, M. Comparative Multiple Regression Analysis of Household Electricity use in Latvia: Using Smart Meter Data to Examine the Effect of Different Household Characteristics. *Energy Procedia* **2015**, *72*, 49–56. [CrossRef]

41. Xie, C.; Hawkes, A.D. Estimation of inter-fuel substitution possibilities in China's transport industry using ridge regression. *Energy* **2015**, *88*, 260–267. [CrossRef]

energies

MDPI

Article

A Carbon Price Forecasting Model Based on Variational Mode Decomposition and Spiking Neural Networks

Guoqiang Sun *, Tong Chen *, Zhinong Wei, Yonghui Sun, Haixiang Zang and Sheng Chen

College of Energy and Electrical Engineering, Hohai University, Nanjing 211100, China; wzn_nj@263.net (Z.W.); sunyonghui168@gmail.com (Y.S.); zanghaixiang@hhu.edu.cn (H.Z.); chenshenghhu@163.com (S.C.)
* Correspondence: hhusunguoqiang@163.com (G.S.); hhuchentong@163.com (T.C.);
   Tel.: +86-136-0514-5395 (G.S.); +86-157-5187-1735 (T.C.)

Academic Editor: Wei-Chiang Hong
Received: 18 November 2015; Accepted: 11 January 2016; Published: 19 January 2016

Abstract: Accurate forecasting of carbon price is important and fundamental for anticipating the changing trends of the energy market, and, thus, to provide a valid reference for establishing power industry policy. However, carbon price forecasting is complicated owing to the nonlinear and non-stationary characteristics of carbon prices. In this paper, a combined forecasting model based on variational mode decomposition (VMD) and spiking neural networks (SNNs) is proposed. An original carbon price series is firstly decomposed into a series of relatively stable components through VMD to simplify the interference and coupling across characteristic information of different scales in the data. Then, a SNN forecasting model is built for each component, and the partial autocorrelation function (PACF) is used to determine the input variables for each SNN model. The final forecasting result for the original carbon price can be obtained by aggregating the forecasting results of all the components. Actual InterContinental Exchange (ICE) carbon price data is used for simulation, and comprehensive evaluation criteria are proposed for quantitative error evaluation. Simulation results and analysis suggest that the proposed VMD-SNN forecasting model outperforms conventional models in terms of forecasting accuracy and reliability.

Keywords: carbon price forecasting; variational mode decomposition (VMD); spiking neural network (SNN); partial autocorrelation function (PACF); comprehensive evaluation criteria

1. Introduction

Global warming induced by fossil fuel consumption has become a formidable challenge faced by all countries, which has spurred the development of socialized economies. Correspondingly, a worldwide consensus has emerged to foster a clean, high efficiency and low-carbon energy system. The Kyoto Protocol agreement officially came into force in 2005, marking the beginning of greenhouse gas reduction by leveraging a market mechanism [1]. Henceforth, the carbon trading market has expanded worldwide. As one of the primary sources of carbon emissions, the power industry has a dramatic potential for carbon emissions reduction and obvious scope for optimization. In the past two decades, numerous studies have been conducted regarding power system planning and dispatching under market circumstances that take into account carbon trading and, in particular, future carbon prices [2–4]. Therefore, the development of a reliable carbon price forecasting and analysis approach is the key to anticipating the changing trends of the energy market, and, thus, to provide a valid reference for establishing power industry policy.

Recently, carbon price forecasting has attracted considerable worldwide attention. Generally speaking, existing models and methods that have been adopted can be mainly divided into two categories: single models [5–7] and combined models [8–10]. Single model forecasting mainly uses generalized autoregressive conditional heteroskedasticity (GARCH)-type models or artificial neural networks (ANNs) to analyze and simulate a carbon price time series, and then employs the developed model to forecast the carbon price. For example, various types of GARCH models, including GARCH, EGARCH and TGARCH, have been proposed [5] to forecast and analyze the volatility of European Union allowance (EUA) spot and futures. The forecasting effect of single variable and multivariable GARCH models in the energy market has also been evaluated [6]. These types of GARCH models are statistical models that are not able to capture the nonlinear characteristics of the carbon price time series effectively, which affects the forecasting accuracy. Compared with statistical models, ANNs possess strong self-learning and adaptive capabilities, and can perform complex nonlinear mapping [7]. Nevertheless, ANNs face substantial challenges in dealing with large historical data sets, which also limits forecasting accuracy. Spiking neural networks (SNNs) are third generation neural networks that use the temporal encoding scheme to transmit information and perform calculations [11], and SNNs have been shown to more realistically reflect the behavior of actual biological nervous systems [12]. Furthermore, SNNs have demonstrated the capability of simulating the function of any feedforward sigmoidal neural network and to approximate arbitrary continuous function [13,14]. As a result, SNNs have exhibited stronger computing capabilities and higher forecasting accuracies than any other type of neural networks, and have been successfully applied to engineering and forecasting fields [15–19].

On the other hand, carbon price series have strong nonlinear and non-stationary characteristics [20,21]. As such, no perfect single model can be applied for accurate forecasting. Therefore, combined forecasting models have integrated empirical mode decomposition (EMD), which is an adaptive signal decomposition algorithm [22–24], and conventional forecasting methods to forecast and analyze the carbon price. As an example of a combined model, a carbon price series was decomposed into a series of relatively stable components through EMD prior to analysis [8], which was found to forecast the underlying characteristics of the carbon price. The same method has been adopted to simplify the interference and coupling across the characteristic information of different scales in the carbon price data [9,10]. Thus, a forecasting model can better infer the characteristics of each component so as to improve the forecasting accuracy. However, EMD is a recursive mode decomposition algorithm, and is limited by mode aliasing and an inability to correctly separate components of similar frequencies [25,26]. These limitations may affect the final forecasting accuracy of the carbon price. To alleviate the deficiencies of the EMD algorithm, Dragomiretskiy *et al.* [25] proposed a new adaptive signal decomposition estimation methodology in 2014, denoted as variational mode decomposition (VMD). Compared with the recursive screening mode of EMD, the VMD algorithm transforms the signal decomposition into a non-recursive and variational model based on a solid theoretical foundation. Thus, VMD demonstrates better noise robustness and more precise component separation [26]. At present, this method has been successfully applied to solve various problems such as international stock market analysis, classification of power quality disturbances and rub-impact fault detection of a rotor system [27–29].

Considering the outstanding advantages of VMD in nonlinear and non-stationary signal decomposition and the superior performance of SNNs in forecasting, as discussed above, a carbon price forecasting model based on VMD and SNNs is proposed in this paper. First, an original carbon price series is decomposed using the VMD algorithm to capture its complicated intrinsic linear and nonlinear characteristics more accurately. Next, the partial autocorrelation function (PACF) [30,31] and the resulting partial autocorrelation graph are employed as statistical tools to determine the input variables of each component. SNNs are then used to build forecasting models for each component to improve the forecasting accuracy. Finally, comprehensive error evaluation criteria comprised of two types of evaluation indexes, including level and phase errors [32], are proposed. The criteria can provide a comprehensive evaluation of the average level and distribution of the forecasting error.

The remainder of this paper is organized as follows: Section 2 describes the processes of the VMD technique and the SNN model. Section 3 elaborates on the combined VMD-SNN forecasting model, which serves as the kernel of this paper. Section 4 presents the comprehensive evaluation criteria and analysis of the simulation results is presented to verify the feasibility and forecasting performance of the proposed model based on the established evaluation criteria. Finally, Section 5 presents the conclusions of the work.

2. Methodologies

2.1. Variational Mode Decomposition (VMD)

VMD transfers the signal decomposition process into a process of solving a variational model to obtain the sub-signals (modes). It is superior to EMD in getting rid of the cycling screening signal processing method. Assuming that each mode has a limited bandwidth with a unique center frequency in the frequency domain, the signal can be adaptively decomposed by obtaining the optimal solution of the constrained variational model. The center frequency and bandwidth of each mode is constantly updated during the variational model solution process. Each mode is demodulated to its corresponding baseband, and, ultimately, all the modes and their corresponding center frequencies are extracted [25]. In the following, the process of the VMD algorithm is described briefly, and the concrete procedures involved are illustrated in Figure 1.

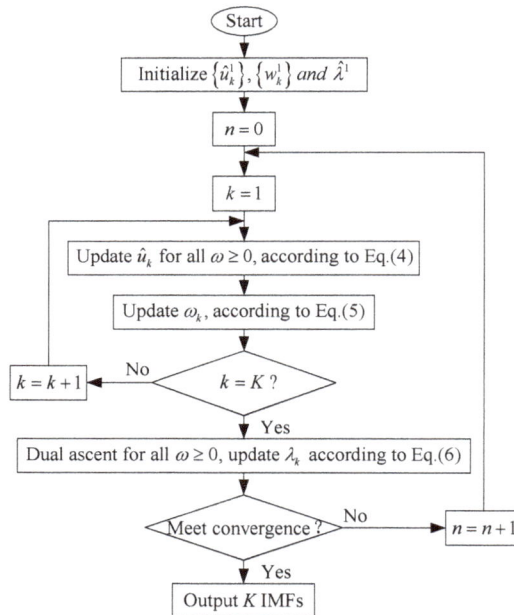

Figure 1. The concrete procedures of variational mode decomposition (VMD).

In the VMD algorithm, the intrinsic mode functions (IMFs) are redefined as amplitude modulated and frequency modulated (AM-FM) signals [25], which are given as a function of time t:

$$u_k(t) = A_k(t)\cos(\phi_k(t)) \tag{1}$$

where $\phi_k(t)$ is the phase, $A_k(t)$ and $w_k(t)$ are the envelope and instantaneous frequency of the kth mode $u_k(t)$, respectively, and $w_k(t) = \phi\prime_k(t)$. Both $A_k(t)$ and $w_k(t)$ vary much more slowly than $\phi_k(t)$, which implies that, on a sufficiently long interval $[t - \delta, t + \delta]$ ($\delta \approx 2\pi/\phi\prime_k(t)$), $u_k(t)$ can be considered to be a pure harmonic signal with amplitude $A_k(t)$ and instantaneous frequency $w_k(t)$.

Assume the original signal f is decomposed into K discrete modes, such that the variational problem can be described as a constrained variational problem. The problem is targeted at minimizing the sum of the estimated bandwidth for each mode by finding K mode functions $u_k(t)$. Then the constrained variational formulation is given as follows:

$$\left. \min_{\{u_k\},\{\omega_k\}} \left\{ \sum_k || \partial_t[(\sigma(t) + \frac{j}{\pi t}) * u_k(t)]e^{-j\omega_k t} ||_2^2 \right\} \atop \text{s.t.} \quad \sum_k u_k = f \right\} \tag{2}$$

Here $\{u_k\} = \{u_1, \dots u_K\}$ and $\{\omega_k\} = \{\omega_1, \dots \omega_K\}$ are the set of all modes and their center frequencies, respectively. $\sigma(t)$ is the Dirac distribution, $j^2 = -1$ and $*$ denotes convolution.

The above constrained variational problem can be addressed by introducing a quadratic penalty factor α and Lagrange multipliers $\lambda(t)$. Therefore, the augmented Lagrangian (L) is formulated as:

$$L(\{u_k\},\{\omega_k\},\lambda) = \alpha \sum_k || \partial_t[(\sigma(t) + \frac{j}{\pi t}) * u_k(t)]e^{-j\omega_k t} ||^2 + ||f(t) - \sum_k u_k(t)||_2^2 + \left\langle \lambda(t), f(t) - \sum_k u_k(t) \right\rangle \tag{3}$$

where $||\bullet||_p$ denotes the usual vector ℓ_p norm ($p = 2$). The optimization methodology denoted as the alternate direction method of multipliers (ADMM) is then used to obtain the saddle point of the augmented Lagrangian by updating u_k^{n+1}, ω_k^{n+1} and λ_k^{n+1} alternately. The convergence criterion of the algorithm is $\sum_k (|| \hat{u}_k^{n+1} - \hat{u}_k^n ||_2^2 / || \hat{u}_k^n ||_2^2) < \varepsilon$, where ε is the convergence tolerance and $\hat{}$ denotes the Fuorier transforms. The final updated equations are given as follows [25]:

$$\hat{u}_k^{n+1}(\omega) = \frac{\hat{f}(\omega) - \sum_{i<k} \hat{u}_i^{n+1}(\omega) - \sum_{i>k} \hat{u}_i^n(\omega) + \frac{\hat{\lambda}^n(\omega)}{2}}{1 + 2\alpha(\omega - \omega_k^n)^2} \tag{4}$$

$$\omega_k^{n+1} = \frac{\int_0^\infty \omega \left| \hat{u}_k^{n+1}(\omega) \right|^2 d\omega}{\int_0^\infty \left| \hat{u}_k^{n+1}(\omega) \right|^2 d\omega} \tag{5}$$

$$\hat{\lambda}^{n+1}(\omega) = \hat{\lambda}^n(\omega) + \tau[\hat{f}(\omega) - \sum_k \hat{u}_k^{n+1}(\omega)] \tag{6}$$

where n is the iteration number and τ is the time step of the dual ascent.

2.2. Spiking Neural Networks (SNNs)

The architecture of a SNN consists of a feedforward network of spiking neurons with multiple delayed synaptic terminals. A spiking neuron is the basic unit of a SNN, and common models of spiking neurons are the leaky integrate-and-fire model (LIFM), Hodgkin-Huxley model (HHM) and spike response model (SRM) [12]. The standard three-layer feedforward SNN based on SRM is adopted in this paper. Figure 2 illustrates the connectivity between an arbitrary spiking neuron h and the ith neuron in successive layers.

From Figure 2, we can see that the SNN differs from conventional back propagation (BP) neural network in that an individual connection consists of a fixed number of m synaptic terminals. Each terminal serves as a sub-connection associated with an adjustable delay d^k and weight W_{hi}^k. A temporal encoding scheme is adopted in SNN, which takes the firing time of a spiking neuron as input and output signals directly. The characteristics of multiple delayed synaptic terminals and the time encoding scheme provide SNNs with not only a powerful computing capability and good applicability, but also time series tractability in particular.

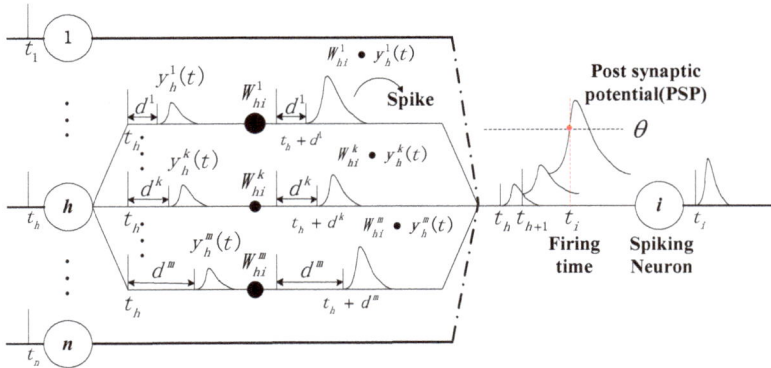

Figure 2. The connectivity between neuron h and the ith neuron with m delayed synaptic terminals.

Each neuron fires at most a single spike during the simulation interval, and only fires when an internal state variable, denoted as the membrane potential (MP), reaches the preset neuronal excitation threshold θ (in Figure 2). Meanwhile, the neuron firing the spike generates an output signal, denoted as the post synaptic potential (PSP), which is defined by the spike-response function $\varepsilon(t)$:

$$\varepsilon(t) = \begin{cases} \dfrac{t}{\tau_s} e^{1-\frac{t}{\tau_s}} & t > 0 \\ 0 & t \leqslant 0 \end{cases} \tag{7}$$

Here, τ_s is the membrane potential decay time constant that determines the rise and decay time of the PSP. Neuron h receives a series of spikes, and its firing time is t_h. Therefore, the actual firing time t_i^a of neuron i is calculated as follows:

$$Y_h^k(t) = \varepsilon(t - t_h - d^k) \tag{8}$$

$$X_i(t) = \sum_h \sum_{k=1}^{m} W_{hi}^k Y_h^k(t) \tag{9}$$

$$t_i^a : X_i(t_i^a) = \theta \ and \ \left.\frac{dX_i(t)}{dt}\right|_{t=t_i^a} > 0 \tag{10}$$

Here, $Y_h^k(t)$ is the unweight contribution of a single synaptic terminal to MP, and $X_i(t)$ is the MP of neuron i. d^k and W_{hi}^k denote the delay and weight associated with k-th synaptic terminal, respectively.

The training algorithm employed for the SNN is spike propagation (SpikeProp), which is a supervised learning algorithm proposed by Bohte *et al.* [14]. In the SpikeProp algorithm, the weight of the network is adjusted to minimize the training error E for the entire network until it is within an established tolerance. A detailed description of the SpikeProp algorithm for a three-layer feedforward SNN is as follows:

(1) Calculate E for the entire network according to the difference between the actual spike firing time t_j^a and the desired spike firing time t_j^d of all neurons j in output layer J, respectively:

$$E = \frac{1}{2} \sum_{j \in J} \left(t_j^a - t_j^d \right)^2 \tag{11}$$

(2) Calculate δ_j for all neurons j in output layer J:

$$\delta_j = \frac{(t_j^d - t_j^a)}{\displaystyle\sum_{i \in \Gamma_j} \sum_k W_{ij}^k \frac{\partial Y_i^k(t_j^a)}{\partial t_j^a}} \tag{12}$$

where Γ_j represents the set of all presynaptic neurons for neuron j.

(3) Calculate δ_i for all neurons i in hidden layer I:

$$\delta_i = \frac{\displaystyle\sum_{j \in \Gamma^i} \delta_j \left\{ \sum_k W_{ij}^k \frac{\partial Y_i^k(t_j^a)}{\partial t_i^a} \right\}}{\displaystyle\sum_{h \in \Gamma_i} \sum_k W_{hi}^k \frac{\partial Y_h^k(t_i^a)}{\partial t_i^a}} \tag{13}$$

where Γ^i and Γ_i are the set of all postsynaptic neurons and presynaptic neurons for neuron i, respectively.

(4) Adjust the weights ΔW_{ij}^k and ΔW_{hi}^k for output layer J and hidden layer I, respectively, according to the network learning rate η as follows.

$$\Delta W_{ij}^k = -\eta \delta_j Y_i^k \left(t_j^a \right) \tag{14}$$

$$\Delta W_{hi}^k = -\eta \delta_i Y_h^k \left(t_j^a \right) \tag{15}$$

3. The Proposed VMD-SNN Forecasting Model

To capture the complicated intrinsic nonlinear and non-stationary characteristics of carbon price data, by using the VMD algorithm, an original carbon price series is firstly decomposed into a series of IMF components with higher regularity than the original data. For the superior performance of SNNs in forecasting, the SNN forecasting model for each stationary IMF component is built by considering the characteristics of that component. To reflect the relationship between the input(s) and output(s) of SNNs, the PACF and the resulting partial autocorrelation graph, which is simply the plots of the PACF against the lag length, are used to determine the input variables for the SNN forecasting model corresponding to each IMF [30,31]. Finally, the forecasting results of all the IMF components are aggregated to produce a combined forecasting result for the original carbon price. A detailed description regarding the determination of input variables is given in Section 3.1, and the overall forecasting procedures are discussed in Section 3.2.

3.1. Determination of Input Variables by PACF

Assuming that x_t is the output variable, if the partial autocorrelation at lag k is outside of the 95% confidence interval $[-1.96/\sqrt{N}, 1.96/\sqrt{N}]$, x_{t-k} is one of the input variables. The previous value x_{t-1} is taken as the input variable if all the PACF coefficients are within the 95% confidence interval. The PACF is described as follows: for a carbon price series $\{x_1, x_2, \cdots x_n\}$, the covariance at lag k (if $k = 0$, it is the variance), denoted by γ_k, is estimated as:

$$\hat{\gamma}_k = \frac{1}{n} \sum_{t=1}^{n-k} (x_t - \bar{x})(x_{t+k} - \bar{x}), \ k = 1, 2, \dots, M \tag{16}$$

where $\bar{x}$ is the mean of the carbon price series and $M = n/4$ is the maximum lag. Then, the autocorrelation function (ACF) at lag k, denoted by ρ_k, can be estimated as:

$$\hat{\rho}_k = \frac{\hat{\gamma}_k}{\hat{\gamma}_0} \tag{17}$$

Based on the covariance and the resulting ACF, the calculation for the PACF at lag k, for $k = 1, 2, \ldots, M$, denoted by α_{kk}, is presented as follows:

$$\left.\begin{array}{l} \hat{\alpha}_{11} = \hat{\rho}_1 \\[2mm] \hat{\alpha}_{k+1,k+1} = \dfrac{\hat{\rho}_{k+1} - \sum_{j=1}^{k} \hat{\rho}_{k+1-j}\hat{\alpha}_{kj}}{1 - \sum_{j=1}^{k} \hat{\rho}_j\hat{\alpha}_{kj}} \\[4mm] \hat{\alpha}_{k+1,j} = \hat{\alpha}_{kj} - \hat{\alpha}_{k+1,k+1} \cdot \hat{\alpha}_{k,k-j+1}, \; j = 1, 2, \ldots, k \end{array}\right\} \tag{18}$$

3.2. Overall Procedures of the VMD-SNN Forecasting Model

The overall procedures of the proposed VMD-SNN forecasting model are illustrated in Figure 3, which are generally comprised of the following main steps:

Step 1. Apply the VMD algorithm to decompose the original carbon price series into K IMF components (sub-series).

Step 2. For each IMF component, with the output variable x_t, the input variables are determined through observing the partial autocorrelogram via PACF.

Step 3. A three-layer SNN forecasting model is built for each IMF component. Perform SNN training using the training sample prior to importing the test sample into the well-trained SNN model. The output is then the forecasting value of the current IMF component.

Step 4. Aggregate the forecasting results of all the IMF components obtained by the previous steps to produce a combined forecasting result for the original carbon price series.

Step 5. The comprehensive error evaluation criteria proposed in this paper are applied to evaluate and analyze the final forecasting result.

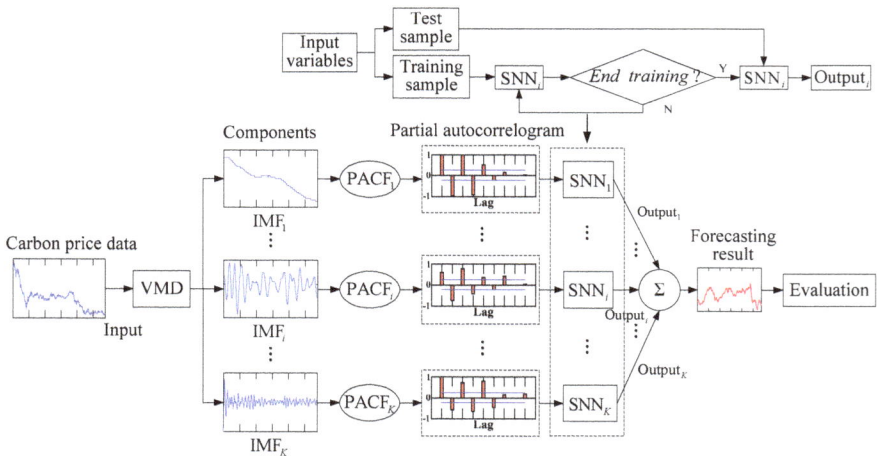

Figure 3. The overall procedures of the VMD-SNN forecasting model.

4. Simulation and Results Analysis

4.1. Data Description

The InterContinental Exchange (ICE) is the largest carbon emissions futures exchange in Europe. The EUA traded in the European Union emissions trading system (EU ETS) has also been leading the world carbon price, and is becoming a representative index of the general carbon price [21]. Therefore, forecasting the price of EUA carbon emissions futures matured in December, 2012 (DEC12) from the ICE can effectively reflect the overall state of the EU ETS at that time. The carbon emissions price data used in this paper are daily transaction data available from the ICE website [33].

Owing to its availability and continuity, the DEC12 carbon price series obtained from 13 June 2008 to 17 December 2012, excluding public holidays, with 1149 total data points, was chosen as experimental samples in this study. For modeling convenience, the data obtained from 13 June 2008 to 22 May 2012, excluding public holidays, with a total of 1000 data points were used as training sets, and the remaining 149 data points were used as testing sets to verify the effectiveness of the VMD-SNN forecasting model based on the comprehensive error evaluation criteria.

In addition, to improve the generalization capability and forecasting accuracy of the network, all the original carbon series and decomposed sub-series must be normalized prior to training and testing as follows:

$$p^* = \frac{p - p_{\min}}{p_{\max} - p_{\min}} \tag{19}$$

Here, p^* represents the sample data after normalization and p is the original sample data. $p_{\max}$ and $p_{\min}$ are the maximum and minimum values of p, respectively.

4.2. Comprehensive Evaluation Criteria

To evaluate the forecasting performance of the proposed VMD-SNN carbon price forecasting model comprehensively and effectively from different perspectives, this paper puts forward comprehensive evaluation criteria taking into account both level and phase errors. The level errors mainly describe the differences between actual and forecasting results in the vertical direction over a given time period. It can be often summed up as smaller or larger. The phase errors mainly describe the differences between actual and forecasting results along the horizontal timeline, which, broadly speaking reflect the leading or lagging between forecasting and actual carbon price series peaks/valleys.

4.2.1. Evaluation Indexes

Assuming that y_i and $\widehat{y}_i$ are the actual and forecasting carbon price values at time i, respectively, and N is the number of data. The root mean square error (*RMSE*), calculated according to Equation (20), measures the degree of forecasting error dispersion, and the mean absolute error (*MAE*), calculated according to Equation (21), measures the average amplitude of error. Combining the two indexes can provide a macroscopic evaluation of the level error characteristics of the forecasting carbon price series. Next, the maximum absolute percentage error (*MaxAPE*), calculated according to Equation (22), concentrates on the maximal absolute percentage error, which reflects the forecasting risk of choosing a particular model:

$$RMSE = \sqrt{\frac{1}{N} \sum_{i=1}^{N} \left(y_i - \widehat{y}_i \right)^2}, \ i = 1, 2, \cdots, N \tag{20}$$

$$MAE = \frac{1}{N} \sum_{i=1}^{N} \left| y_i - \widehat{y}_i \right|, \ i = 1, 2, \cdots, N \tag{21}$$

$$MaxAPE = \max_i \left(\left| \frac{y_i - \widehat{y}_i}{y_i} \right| \right) \times 100, \ i = 1, 2, \cdots, N \tag{22}$$

The correlation coefficient denoted by I_{cc}, which primarily measures the level and phase errors directly, can describe the correlation between the actual and forecasting carbon price series. The forecasting accuracy increases with increasing I_{cc}, and is defined as follows:

$$I_{cc} = \frac{\text{cov}(Y_a, Y_f)}{\sqrt{DY_a}\sqrt{DY_f}} \tag{23}$$

where Y_a and Y_f are the actual and forecasting carbon price series, respectively, and D represents the variance.

4.2.2. Histogram of the Error Frequency Distribution

This index is an improvement of the mean error (ME), which substitutes the original averaging process for the form of frequency distribution histogram. This index reserves the function of ME to measure whether the system is unbiased, in addition, it gives distribution of the error bands in the forecasting result. The degree of concentration for the error frequency distribution relative to zero can be regarded as a basis for comparing different forecasting models.

4.3. Parameter Setting

To demonstrate the effectiveness of the proposed VMD-SNN forecasting model, the single BP and SNN models are employed as benchmarks for comparison. In addition, the combined EMD-SNN and VMD-BP models are also used to forecast the carbon price for the purpose of comparison. The forecasting simulations are performed on a MATLAB® (R2014a) platform, and the simulation parameters are set as follows.

The number of modes K should be firstly determined when applying the VMD algorithm to decompose the original DEC12 carbon price series. The primary distinguishing characteristic of a mode is its center frequency. Thus, for simplicity, we determined K on the basis of the method's capability to distinguish center frequencies. For DEC12, modes with similar center frequencies are observed to appear when K reaches 6, and we argue that this corresponds with the onset of greedy decomposition. Consequently, K is set to 6. The penalty factor α and the tolerance of convergence criterion ε are set to default values of 2000 and 10^{-6}, respectively. Moreover, the noise with low level in the original data can be filtered by adjusting the update parameter $\tau > 0$, and the value is set to 0.3 in this paper. For the EMD algorithm, the thresholds and tolerance level of the stop criterion are determined as $[\theta_1, \theta_1, \alpha] = [0.05, 0.5, 5 \times 10^{-4}]$.

In the SNN model, the number of synaptic terminals m is set to 16, and the corresponding synapses delay d^k is selected as an incremental integer in the interval $[1, 16]$. The decay time constant τ_s is set to 6 ms, and the excitation threshold θ for all neurons is set as 1 mV. For both SNN and BP, the maximum iterations and learning rate are set as 1000 and 0.0005, respectively.

4.4. Results and Analysis

The forecasting simulations are conducted according to the procedures discussed in Section 3.2. Firstly, the proposed VMD algorithm and the widely used EMD algorithm are adopted to decompose the DEC12 carbon price series, and the decomposition results are illustrated in Figures 4 and 5 respectively.

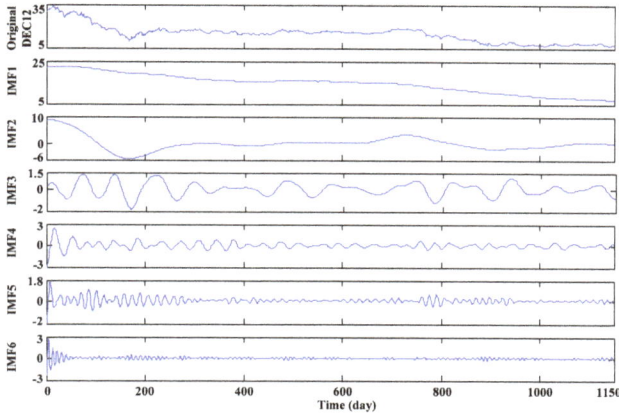

Figure 4. The decomposition of DEC12 using the VMD algorithm.

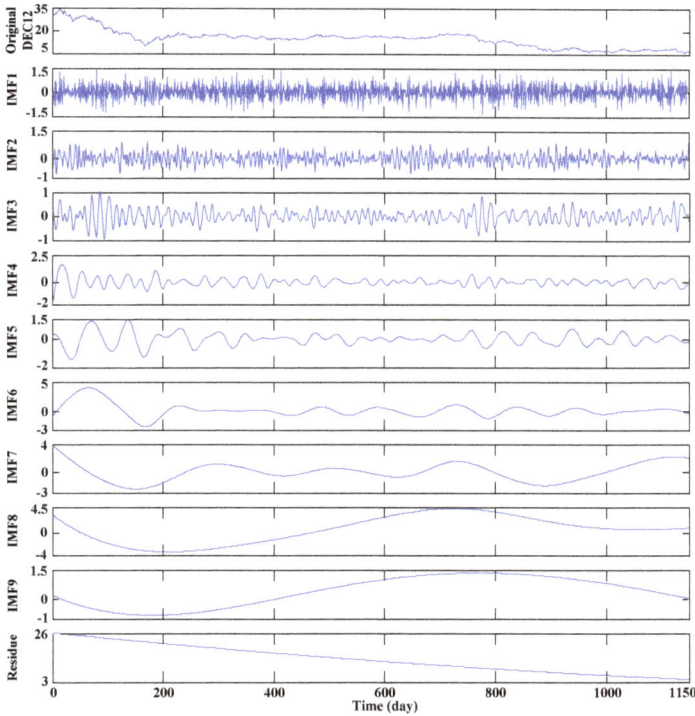

Figure 5. The decomposition of DEC12 using the EMD algorithm.

Based on Figures 4 and 5 it can be observed that the original DEC12 series is decomposed into six IMFs via VMD and nine IMFs plus one residue via EMD. It is noteworthy that, although the carbon price is positive, the values of IMFs may be negative.

The VMD algorithm extracts the significant volatility and trend sub-series at equivalent scales in the original DEC12. In addition, relative to the IMFs derived from VMD, the values of IMF1–IMF3 derived from EMD oscillate with very high frequencies, which results in additional difficulties for forecasting. Then, using the PACF, the partial autocorrelogram of DEC12 and the obtained IMFs can be conveniently derived, which are shown in Figures 6 and 7 respectively. Based on evaluation of the partial autocorrelograms, the input variables of the original carbon price series and the IMFs obtained via the two decomposition algorithms corresponding to output variable x_t are presented in Table 1.

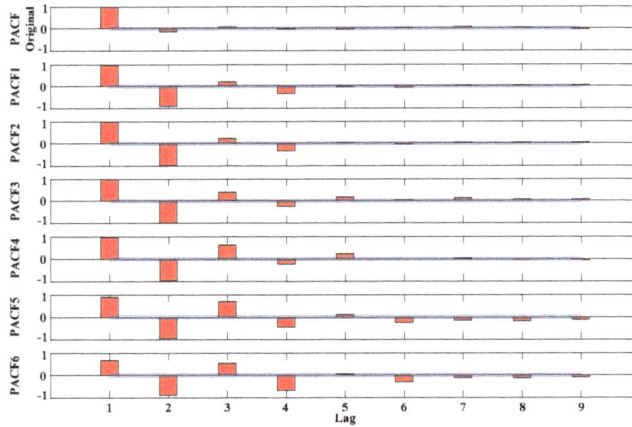

Figure 6. The PACFs of the original series and the IMFs derived from VMD.

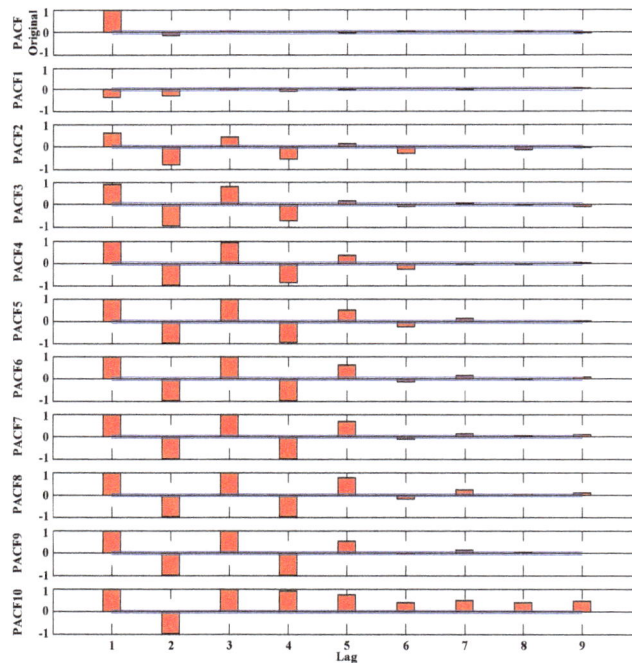

Figure 7. The PACFs of the original series and the IMFs derived from EMD.

Table 1. Input variables of the original series and the IMFs.

Series	Mode Decomposition Algorithm	
	VMD	EMD
DEC12	$(x_{t-1}, x_{t-2}, x_{t-3}, x_{t-7})$	$(x_{t-1}, x_{t-2}, x_{t-3}, x_{t-7})$
IMF1	$(x_{t-1}, x_{t-2}, x_{t-3}, x_{t-4}, x_{t-6})$	$(x_{t-1}, x_{t-2}, x_{t-3}, x_{t-4})$
IMF2	$(x_{t-1}, x_{t-2}, x_{t-3}, x_{t-4})$	$(x_{t-1}, x_{t-2}, x_{t-3}, x_{t-4}, x_{t-5}, x_{t-6}, x_{t-8})$
IMF3	$(x_{t-1}, x_{t-2}, x_{t-3}, x_{t-4}, x_{t-5}, x_{t-7}, x_{t-8}, x_{t-9})$	$(x_{t-1}, x_{t-2}, x_{t-3}, x_{t-4}, x_{t-5}, x_{t-6}, x_{t-8}, x_{t-9})$
IMF4	$(x_{t-1}, x_{t-2}, x_{t-3}, x_{t-4}, x_{t-5}, x_{t-9})$	$(x_{t-1}, x_{t-2}, x_{t-3}, x_{t-4}, x_{t-5}, x_{t-6})$
IMF5	$(x_{t-1}, x_{t-2}, x_{t-3}, x_{t-4}, x_{t-5}, x_{t-6}, x_{t-7}, x_{t-8}, x_{t-9})$	$(x_{t-1}, x_{t-2}, x_{t-3}, x_{t-4}, x_{t-5}, x_{t-6}, x_{t-7})$
IMF6	$(x_{t-1}, x_{t-2}, x_{t-3}, x_{t-4}, x_{t-6}, x_{t-7}, x_{t-8}, x_{t-9})$	$(x_{t-1}, x_{t-2}, x_{t-3}, x_{t-4}, x_{t-5}, x_{t-6}, x_{t-7})$
IMF7	—	$(x_{t-1}, x_{t-2}, x_{t-3}, x_{t-4}, x_{t-5}, x_{t-6}, x_{t-7}, x_{t-9})$
IMF8	—	$(x_{t-1}, x_{t-2}, x_{t-3}, x_{t-4}, x_{t-5}, x_{t-6}, x_{t-7}, x_{t-9})$
IMF9	—	$(x_{t-1}, x_{t-2}, x_{t-3}, x_{t-4}, x_{t-5}, x_{t-7}, x_{t-9})$
Residue	—	$(x_{t-1}, x_{t-2}, x_{t-3}, x_{t-4}, x_{t-5}, x_{t-6}, x_{t-7}, x_{t-8}, x_{t-9})$

After determining the input variables, basic single BP and SNN models are used to forecast the original carbon price series and the obtained IMFs. Meanwhile, the forecasting results of the IMFs were aggregated to produce a combined forecasting result for the VMD-based and EMD-based modeling processes, respectively. The actual DEC12 carbon price series and the final forecasting results of various forecasting models are presented in Figure 8. In addition, to facilitate comparison, enlarged views of the carbon price series curves shown in Figure 8 in the time intervals reflecting strong volatility are given in Figure 9. Enlarged views basically cover the peaks and valleys of the actual DEC12 carbon price series curve, and can reflect the fluctuation characteristics of the DEC12 and the tracking performance of forecasting models.

It follows from the DEC12 data shown in Figure 8 that, despite the periodic volatility at different scales, an abnormal stochastic volatility is observed in the carbon price series. Therefore, the complexity of the carbon price variation makes accurate forecasting difficult. Nevertheless, the forecasting result of the proposed VMD-SNN model closely tracks the changing trends of the actual carbon price, even nearby inflection points, which is made particularly apparent from Figure 9 under conditions of extreme price volatility. The tracking performances of the other forecasting models considered in this paper are obviously inferior. For example, the forecasting results of single BP and SNN models exhibit large level errors in time interval [1,18], [111,128], [45,55] and [135,145]. On the other hand, large phase errors exist in the forecasting results of the combined EMD-SNN and VMD-BP models in time interval [135,145].

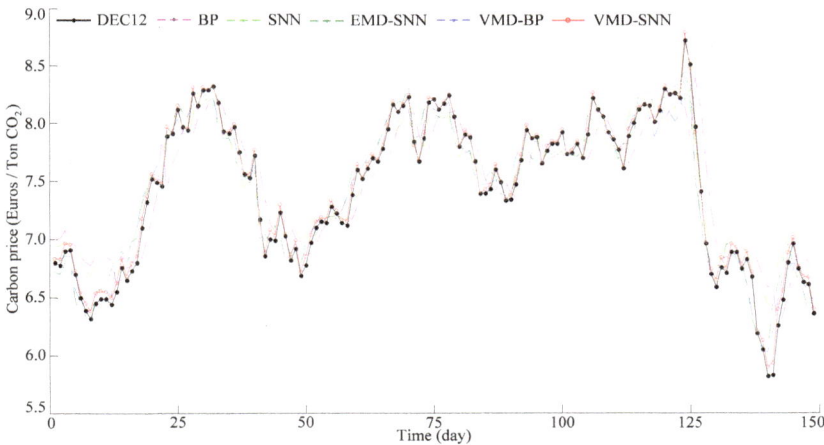

Figure 8. Actual carbon price and the forecasting results of various models.

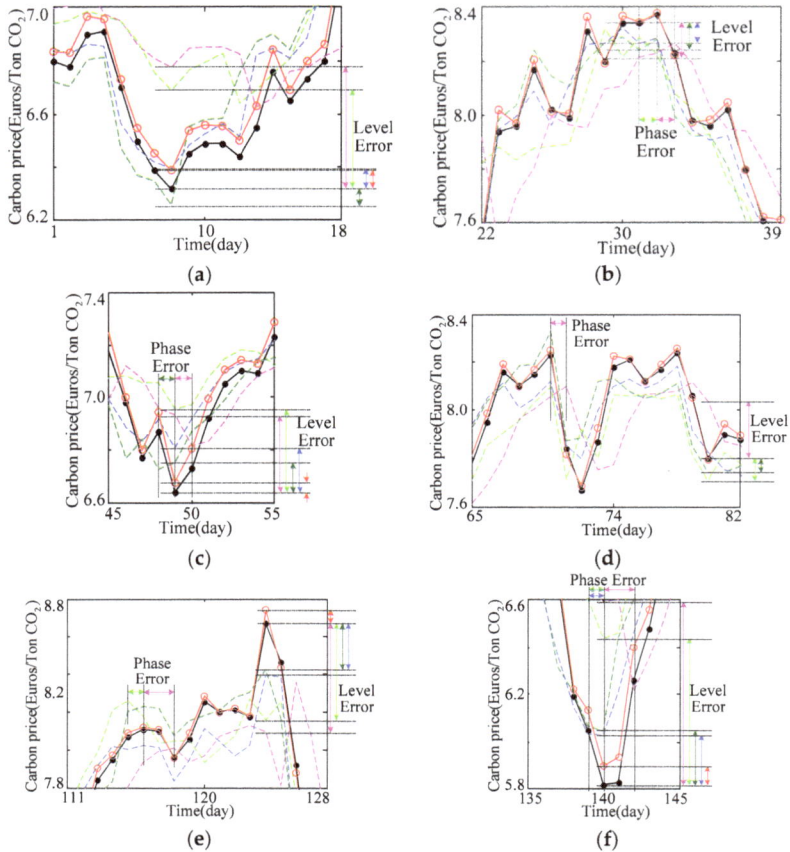

Figure 9. Enlarged views of the carbon price series curves in the time interval: (**a**) [1,18]; (**b**) [22,39]; (**c**) [45,55]; (**d**) [65,82]; (**e**) [111,128]; (**f**) [135,145].

For a quantitative analysis of the forecasting results, Table 2 lists the error statistics for the various forecasting models considered. In Table 2, the *RMSE, MAE, MaxAPE* and I_{cc} values of the VMD-SNN model are 0.0437, 0.0355, 2.197% and 0.9993, respectively, indicating that the forecasting accuracy is higher and the forecasting risk is smaller when using the proposed VMD-SNN model to forecast the carbon price. Based on Table 2, we can further conclude the following: (a) The forecasting performance of the single SNN model is better than that of the conventional BP model, which indicates that SNN is more capable of strong nonlinear mapping, and is more applicable to forecasting the dynamic and nonlinear characteristics of the carbon price; (b) The forecasting performances of the combined forecasting models are superior to any single model. This may be attributable to the fact that mode decomposition can capture the complicated intrinsic characteristics, including both linear and nonlinear characteristics, in the carbon price series. Therefore, integrating a mode decomposition method within forecasting models has a positive effect on the overall forecasting capability; (c) That the proposed VMD-SNN forecasting model outperforms the EMD-SNN model can be mainly attributed to the fact that the carbon price series is decomposed more accurately via the VMD algorithm. This emphasizes the significance of the VMD algorithm for improving forecasting performance in carbon price forecasting.

Table 2. Error statistics for various forecasting models.

Evaluation Indexes	Forecasting Models				
	BP	**SNN**	**EMD-SNN**	**VMD-BP**	**VMD-SNN**
RMSE	0.2655	0.2077	0.1528	0.1231	**0.0437**
MAE	0.2062	0.1690	0.1220	0.0955	**0.0355**
MaxAPE (%)	13.239	10.827	10.546	6.495	**2.197**
I_{cc}	0.9180	0.9621	0.9709	0.9822	**0.9993**

Figure 10 shows histograms of the error frequency distribution for the various forecasting models considered. The frequency with which the forecasting error $e_{ME}\%$ of the proposed VMD-SNN model resides within the interval (0, 10%) is approximately 90%. It is obvious that the range of error distribution of the proposed VMD-SNN model is very narrow, and the degree of concentration for the error frequency distribution relative to zero is much greater than that of the other forecasting models. Moreover, the limit error of the VMD-SNN model is slightly greater than 20%, which is far smaller than that of the other models. The above analysis implies that the proposed VMD-SNN model yields significant improvements in carbon price forecasting.

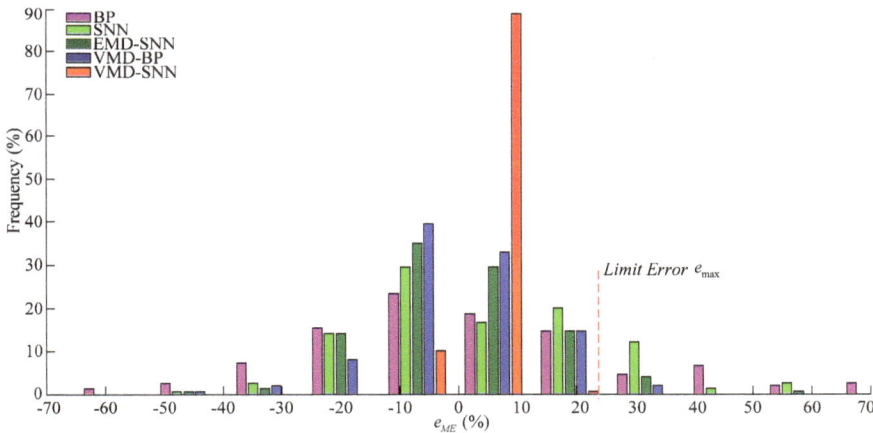

Figure 10. Histograms of the error frequency distribution for the various forecasting models.

5. Conclusions

Carbon price forecasting is a difficult and complex task owing to the non-linear and non-stationary characteristics of carbon price series data. It is scarcely possible to achieve a satisfactory forecasting result by simply employing a single model. This paper focused on aggregating multi-algorithms to exploit the intrinsic characteristics of carbon price series data, and a VMD-SNN combined forecasting model was proposed to forecast the carbon price. Using VMD, the carbon price series is decomposed into volatility and trend sub-series at equivalent scales, such that the complicated intrinsic linear and nonlinear characteristics are appropriately captured. The SNN model, which is capable of strong nonlinear mapping, is proposed as a forecaster for each decomposition component. Additionally, the PACF and the partial autocorrelogram are used as statistical tools to determine the input variables for the resulting SNN forecasting models. The individual results are aggregated to produce a final forecasting result for the original carbon price series. The actual daily transaction price data of EUA from the ICE obtained from 13 June 2008 to 22 May 2012 is used to verify the effectiveness of the proposed VMD-SNN forecasting model, and quantitative evaluation is conducted based on comprehensive error evaluation criteria. Simulation results and analysis demonstrate that the influence of nonlinear and non-stationary characteristics of data on the carbon price forecasting is reduced by the VMD-SNN model. Therefore the proposed VMD-SNN combined forecasting model performs better than conventional single models and EMD-based combined models. The simulation results also suggest that the model not only can be used to forecast the carbon price, but also can be extensively applied to other forecasting fields.

Acknowledgments: The work is supported by the National Natural Science Foundation of China (No. 51277052, 51507052) and China Postdoctoral Science Foundation (2015M571653).

Author Contributions: Guoqiang Sun and Tong Chen are the principal investigators of this work. They proposed the forecasting method, performed the simulations and wrote the manuscript; Zhinong Wei and Yonghui Sun designed the simulations solution and checked the whole manuscript. Haixiang Zang and Sheng Chen contributed to the data analysis work and language editing. All authors revised and approved for the publication.

Conflicts of Interest: The authors declare no conflict of interest.

References

1. Alberola, E.; Chevallier, J.; Chèze, B. Price drivers and structural breaks in European carbon price 2005–2007. *Energy Policy* **2008**, *36*, 787–797. [CrossRef]
2. Liu, Y.; Huang, G.; Cai, Y. An inexact mix-integer two-stage linear programming model for supporting the management of a low-carbon energy system in China. *Energies* **2011**, *4*, 1657–1686. [CrossRef]
3. Yang, J.; Feng, X.; Tang, Y. A power system optimal dispatch strategy considering the flow of carbon emissions and large consumers. *Energies* **2015**, *8*, 9087–9106. [CrossRef]
4. Zhu, Y.; Li, Y.P.; Huang, G.H. A dynamic model to optimize municipal electric power systems by considering carbon emission trading under uncertainty. *Energy* **2015**, *88*, 636–649. [CrossRef]
5. Chevallier, J. Carbon futures and macroeconomic risk factors: A view from the EU ETS. *Energy Econ.* **2009**, *31*, 614–625. [CrossRef]
6. Wang, Y.; Wu, C. Forecasting energy market volatility using GARCH models: Can multivariate models beat univariate models? *Energy Econ.* **2012**, *34*, 2167–2181. [CrossRef]
7. Tsai, M.T.; Kuo, Y.T. A forecasting system of carbon price in the carbon trading markets using artificial neural network. *Int. J. Environ. Sci. Dev.* **2013**, *4*, 163–167. [CrossRef]
8. Zhu, B.; Wang, P.; Chevallier, J. Carbon price analysis using empirical mode decomposition. *Comput. Econ.* **2014**, *45*, 195–206. [CrossRef]
9. Zhu, B. A novel multiscale combined carbon price prediction model integrating empirical mode decomposition, genetic algorithm and artificial neural network. *Energies* **2012**, *5*, 355–370. [CrossRef]
10. Li, W.; Lu, C. The research on setting a unified interval of carbon price benchmark in the national carbon trading market of China. *Appl. Energy* **2015**, *155*, 728–739. [CrossRef]
11. Maass, W. Networks of spiking neurons: The third generation of neural network models. *Neural Netw.* **1997**, *10*, 1659–1671. [CrossRef]
12. Gerstner, W.; Kistler, W.M. *Spiking Neuron Models: Single Neurons, Populations, Plasticity*; Cambridge University Press: Cambridge, UK, 2002.
13. Ponulak, F.; Kasinski, A. Introduction to spiking neural networks: Information processing, learning and applications. *Acta Neurobiol. Exp.* **2010**, *71*, 409–433.
14. Bohte, S.M.; Kok, J.N.; La-Poutre, H. Error-backpropagation in temporally encoded networks of spiking neurons. *Neurocomputing* **2002**, *48*, 17–37. [CrossRef]
15. Rowcliffe, P.; Feng, J. Training spiking neuronal networks with applications in engineering tasks. *IEEE Trans. Neural Netw.* **2008**, *19*, 1626–1640. [CrossRef] [PubMed]
16. Ghosh-Dastidar, S.; Adeli, H. Improved spiking neural networks for EEG classification and epilepsy and seizure detection. *Integr. Comput. Aided Eng.* **2007**, *14*, 187–212.
17. Kulkarni, S.; Simon, S.P.; Sundareswaran, K. A spiking neural network (SNN) forecast engine for short-term electrical load forecasting. *Appl. Soft Comput.* **2013**, *13*, 3628–3635. [CrossRef]

18. Sharma, V.; Srinivasan, D. A spiking neural network based on temporal encoding for electricity price time series forecasting in deregulated markets. In Proceedings of the 2010 IEEE International Joint Conference on Neural Networks (IJCNN), Barcelona, Spain, 18–23 July 2010.

19. Lin, Y.; Teng, Z.J. Prediction of grain yield based on spiking neural networks model. In Proceedings of the 3rd IEEE International Conference on Communication Software and Networks (ICCSN), Xi'an, China, 27–29 May 2011.

20. Seifert, J.; Uhrig-Homburg, M.; Wagner, M. Dynamic behavior of CO2 spot price. *J. Environ. Econ. Manag.* **2008**, *56*, 180–194. [CrossRef]

21. Zhang, Y.J.; Wei, Y.M. An overview of current research on EU ETS: Evidence from its operating mechanism and economic effect. *Appl. Energy* **2010**, *87*, 1804–1814. [CrossRef]

22. Huang, N.E.; Shen, Z.; Long, S.R.; Wu, M.C.; Shih, H.H.; Zheng, Q.; Yen, N.C.; Tung, C.C.; Liu, H.H. The empirical mode decomposition and the Hilbert spectrum for nonlinear and nonstationary time series analysis. *Proc. R. Soc. Lond. A* **1998**, *454*, 903–995. [CrossRef]

23. Hong, Y.Y.; Yu, T.H.; Liu, C.Y. Hour-ahead wind speed and power forecasting using empirical mode decomposition. *Energies* **2013**, *6*, 6137–6152. [CrossRef]

24. Fan, G.F.; Qing, S.; Wang, H. Support vector regression model based on empirical mode decomposition and auto regression for electric load forecasting. *Energies* **2013**, *6*, 1887–1901. [CrossRef]

25. Dragomiretskiy, K.; Zosso, D. Variational mode decomposition. *IEEE Trans. Signal Proc.* **2014**, *62*, 531–544. [CrossRef]

26. Upadhyay, A.; Pachori, R.B. Instantaneous voiced/non-voiced detection in speech signals based on variational mode decomposition. *J. Frankl. Inst.* **2015**, *352*, 2679–2707. [CrossRef]

27. Lahmiri, S. Long memory in international financial markets trends and short movements during 2008 financial crisis based on variational mode decomposition and detrended fluctuation analysis. *Phys. A Stat. Mech. Appl.* **2015**, *437*, 130–138. [CrossRef]

28. Aneesh, C.; Kumar, S.; Hisham, P.M. Performance comparison of variational mode decomposition over empirical wavelet transform for the classification of power quality disturbances using support vector machine. *Procedia Comput. Sci.* **2015**, *46*, 372–380. [CrossRef]

29. Wang, Y.; Markert, R.; Xiang, J. Research on variational mode decomposition and its application in detecting rub-impact fault of the rotor system. *Mech. Syst. Signal Proc.* **2015**, *60*, 243–251. [CrossRef]

30. Dégerine, S.; Lambert-Lacroix, S. Characterization of the partial autocorrelation function of nonstationary time series. *J. Multivar. Anal.* **2003**, *87*, 46–59. [CrossRef]

31. Guo, Z.; Zhao, W.; Lu, H. Multi-step forecasting for wind speed using a modified EMD-based artificial neural network model. *Renew. Energy* **2012**, *37*, 241–249.

32. Giebel, G.; Brownsword, R.; Kariniotakis, G. *The State-of-the-Art in Short-Term Prediction of Wind Power: A Literature Overview*, 2nd ed.; ANEMOS. Plus: Risoe DTU, Roskilde, Denmark, 2011.

33. ICE EUA Futures Contract: Historic Data 2008-2012. Available online: http://www.theice.com (accessed on 24 December 2014).

energies

MDPI

Article

Hybridization of Chaotic Quantum Particle Swarm Optimization with SVR in Electric Demand Forecasting

Min-Liang Huang

Department of Industrial Management, Oriental Institute of Technology/58 Sec. 2, Sichuan Rd, Panchiao, New Taipei 220, Taiwan; fg002@mail.oit.edu.tw; Tel.: +886-2-7738-0145 (ext. 5118); Fax: +886-2-7738-5461

Academic Editor: Sukanta Basu
Received: 25 February 2016; Accepted: 24 May 2016; Published: 31 May 2016

Abstract: In existing forecasting research papers support vector regression with chaotic mapping function and evolutionary algorithms have shown their advantages in terms of forecasting accuracy improvement. However, for classical particle swarm optimization (PSO) algorithms, trapping in local optima results in an earlier standstill of the particles and lost activities, thus, its core drawback is that eventually it produces low forecasting accuracy. To continue exploring possible improvements of the PSO algorithm, such as expanding the search space, this paper applies quantum mechanics to empower each particle to possess quantum behavior, to enlarge its search space, then, to improve the forecasting accuracy. This investigation presents a support vector regression (SVR)-based load forecasting model which hybridizes the chaotic mapping function and quantum particle swarm optimization algorithm with a support vector regression model, namely the SVRCQPSO (support vector regression with chaotic quantum particle swarm optimization) model, to achieve more accurate forecasting performance. Experimental results indicate that the proposed SVRCQPSO model achieves more accurate forecasting results than other alternatives.

Keywords: support vector regression (SVR); chaotic quantum particle swarm optimization (CQPSO); quantum behavior; electric load forecasting

---

1. Introduction

Electric demand forecasting plays the critical role in the daily operational and economic management of power systems, such as energy transfer scheduling, transaction evaluation, unit commitment, fuel allocation, load dispatch, hydrothermal coordination, contingency planning load shedding, and so on [1]. Therefore, a given percentage of forecasting error implies great losses for the utility industries in the increasingly competitive market, as decision makers take advantage of accurate forecasts to make optimal action plans. As mentioned by Bunn and Farmer [2], a 1% increase in electric demand forecasting error represents a £10 million increase in operating costs. Thus, it is essential to improve the forecasting accuracy or to develop new approaches, particularly for those countries with limited energy [3].

In the past decades, many researchers have proposed lots of methodologies to improve electric demand forecasting accuracy, including traditional linear models, such as the ARIMA (auto-regressive integrated moving average) model [4], exponential smoothing models [5], Bayesian estimation model [6], state space and Kalman filtering technologies [7,8], regression models [9], and other time series technologies [10]. Due to the complexity of load forecasting, with these mentioned models it is difficult to illustrate well the nonlinear characteristics among historical data and exogenous factors, and they cannot always achieve satisfactory performance in terms of electric demand forecasting accuracy.

Since the 1980s, due to superior nonlinear mapping ability, the intelligent techniques like expert systems, fuzzy inference, and artificial neural networks (ANNs) [11] have become very successful applications in dealing with electric demand forecasting. In addition, these intelligent approaches can be hybridized to form new novel forecasting models, for example, the random fuzzy variables with ANNs [12], the hybrid Monte Carlo algorithm with the Bayesian neural network [13], adaptive network-based fuzzy inference system with RBF neural network [14], extreme learning machine with hybrid artificial bee colony algorithm [15], fuzzy neural network (WFNN) [16], knowledge-based feedback tuning fuzzy system with multi-layer perceptron artificial neural network (MLPANN) [17], and so on. Due to their multi-layer structure and corresponding outstanding ability to learn non-linear characteristics, ANN models have the ability to achieve more accurate performance of a continuous function described by Kromogol's theorem. However, the main shortcoming of the ANN models are their structure parameter determination [18]. Complete discussions for the load forecasting modeling by ANNs are shown in references [19,20].

Support vector regression (SVR) [21], which has been widely applied in the electric demand forecasting field [11,22–33], hybridizes different evolutionary algorithms with various chaotic mapping functions (logistic function, cat mapping function) to simultaneously and carefully optimize the three parameter combination, to obtain better forecasting performance. As concluded in Hong's series of studies, determination of these three parameters will critically influence the forecasting performance, *i.e.*, low forecasting accuracy (premature convergence and trapped in local optimum) results from the theoretical limitations of the original evolutionary algorithms. Therefore, Hong and his successors have done a series of trials on hybridization of evolutionary algorithms with a SVR model. However, each algorithm has its embedded drawbacks, so to overcome these shortcomings, they continue applying chaotic mapping functions to enrich the searching ergodically over the whole space to do more compact searching in chaotic space, and also apply cloud theory to solve well the decreasing temperature problem during the annealing process to meet the requirement of continuous decrease in actual physical annealing processes, and then, improve the search quality of simulated annealing algorithms, eventually, improving the forecasting accuracy.

Inspired by Hong's efforts mentioned above, the author considers the core drawback of the classical PSO algorithm, which results in an earlier standstill of the particles and loss of activities, eventually causing low forecasting accuracy, therefore, this paper continues to explore possible improvements of the PSO algorithm. As known in the classical PSO algorithm, the particle moving in the search space follows Newtonian dynamics [34], so the particle velocity is always limited, the search process is limited and it cannot cover the entire feasible area. Thus, the PSO algorithm is not guaranteed to converge to the global optimum and may even fail to find local optima. In 2004, Sun *et al.* [35] applied quantum mechanics to propose the quantum delta potential well PSO (QDPSO) algorithm by empowering the particles to have quantum behaviors. In a quantum system, any trajectory of any particles is non-determined, *i.e.*, any particles can appear at any position in the feasible space if it has better fitness value, even far away from the current one. Therefore, this quantum behavior can efficiently enable each particle to expand the search space and to avoid being trapped in local minima. Many improved quantum-behaved swarm optimization methods have been proposed to achieve more satisfactory performance. Davoodi *et al.* [36] proposed an improved quantum-behaved PSO-simplex method (IQPSOS) to solve power system load flow problems; Kamberaj [37] also proposed a quantum-behaved PSO algorithm (q-GSQPO) to forecast the global minimum of potential energy functions; Li *et al.* [38] proposed a dynamic-context cooperative quantum-behaved PSO algorithm by incorporating the context vector with other particles while a cooperation operation is completed. In addition, Coelho [39] proposed an improved quantum-behaved PSO by hybridization with a chaotic mutation operator. However, like the PSO algorithm, the QPSO algorithm still easily suffers from shortcomings in iterative operations, such as premature convergence problems.

In this paper, the author applies quantum mechanics to empower each particle in the PSO algorithm to possess quantum behavior to enlarge the search space, then, a chaotic mapping function is employed to help the particles break away the local optima while the premature condition appears in each iterative searching process, eventually, improving the forecasting accuracy. Finally, the forecasting performance of the proposed hybrid chaotic quantum PSO algorithm with an SVR model, named SVRCQPSO model, is compared with four other existing forecasting approaches proposed in Hong [33] to illustrate its superiority in terms of forecasting accuracy.

This paper is organized as follows: Section 2 illustrates the detailed processes of the proposed SVRCQPSO model. The basic formulation of SVR, the QPSO algorithm, and the CQPSO algorithm will be further introduced. Section 3 employs two numerical examples and conducts the significant comparison among alternatives presented in an existing published paper in terms of forecasting accuracy. Finally, some meaningful conclusions are provided in Section 4.

2. Methodology of SVRCQPSO Model

2.1. Support Vector Regression (SVR) Model

The brief introduction of an SVR model is illustrated as follows. A nonlinear mapping function, $\phi(\cdot)$, is used to map the training data set into a high dimensional feature space. In the feature space, an optimal linear function, f, is theoretically found to formulate the relationship between training fed-in data and fed-out data. This kind of optimal linear function is called SVR function and is shown as Equation (1):

$$f(\mathbf{x}) = \mathbf{w}^{\mathrm{T}}\phi(\mathbf{x}) + b \tag{1}$$

where $f(\mathbf{x})$ denotes the forecasting values; the coefficients $\mathbf{w}$ and b are adjustable. SVR method aims at minimizing the training error, that is the so-called empirical risk, as shown in Equation (2):

$$
\begin{aligned}
R_{emp}(f) &= \frac{1}{N}\sum_{i=1}^{N}\Theta_\varepsilon\left(\mathbf{y}_i, \mathbf{w}^{\mathrm{T}}\phi(\mathbf{x}_i) + b\right) \\
\Theta_\varepsilon(\mathbf{y}, f(\mathbf{x})) &= \begin{cases} |f(\mathbf{x}) - \mathbf{y}| - \varepsilon, & \text{if} \quad |f(\mathbf{x}) - \mathbf{y}| \geqslant \varepsilon \\ 0, & \text{otherwise} \end{cases}
\end{aligned}
\tag{2}
$$

where $\Theta_\varepsilon(\mathbf{y}, f(\mathbf{x}))$ is the ε-insensitive loss function. The ε-insensitive loss function is used to find out an optimum hyper plane on the high dimensional feature space to maximize the distance separating the training data into two subsets. Thus, the SVR focuses on finding the optimum hyperplane and minimizing the training error between the training data and the ε-insensitive loss function. The SVR model then minimizes the overall errors as shown in Equation (3):

$$\underset{\mathbf{w},b,\xi^*,\xi_i}{\text{Min}} \quad R_\varepsilon(\mathbf{w}, \xi^*, \xi) = \frac{1}{2}\mathbf{w}^{\mathrm{T}}\mathbf{w} + C\sum_{i=1}^{N}(\xi_i^* + \xi_i)$$

with the constraints

$$
\begin{aligned}
\mathbf{y}_i - \mathbf{w}^{\mathrm{T}}\phi(\mathbf{x}_i) - b &\leqslant \varepsilon + \xi_i^*, & i = 1, 2, \ldots, N \\
-\mathbf{y}_i + \mathbf{w}^{\mathrm{T}}\phi(\mathbf{x}_i) + b &\leqslant \varepsilon + \xi_i, & i = 1, 2, \ldots, N \\
\xi_i^* &\geqslant 0, & i = 1, 2, \ldots, N \\
\xi_i &\geqslant 0, & i = 1, 2, \ldots, N
\end{aligned}
\tag{3}
$$

The first term of Equation (3), by employed the concept of maximizing the distance of two separated training data, is used to regularize weight sizes, to penalize large weights, and to maintain regression function flatness. The second term, to penalize the training errors of $f(\mathbf{x})$ and $\mathbf{y}$, decides the balance between confidence risk and experience risk by using the ε-insensitive loss function. C is a parameter to trade off these two terms. Training errors above ε are denoted as ξ_i^*, whereas training errors below $-\varepsilon$ are denoted as ξ_i.

After the quadratic optimization problem with inequality constraints is solved, the parameter vector **w** in Equation (1) is obtained with Equation (4):

$$\mathbf{w} = \sum_{i=1}^{N} (\alpha_i^* - \alpha_i)\, \phi(\mathbf{x}_i) \tag{4}$$

where α_i^*, α_i are obtained by solving a quadratic program and are the Lagrangian multipliers. Finally, the SVR regression function is obtained as Equation (5) in the dual space:

$$f(\mathbf{x}) = \sum_{i=1}^{N} (\alpha_i^* - \alpha_i)\, K(\mathbf{x}_i, \mathbf{x}_j) + b \tag{5}$$

where $K(\mathbf{x}_i, \mathbf{x}_j)$ is so-called the kernel function, and the value of the kernel equals the inner product of two vectors, $\mathbf{x}_i$ and $\mathbf{x}_j$, in the feature space $\phi(\mathbf{x}_i)$ and $\phi(\mathbf{x}_j)$, respectively; that is, $K(\mathbf{x}_i, \mathbf{x}_j) = \phi(\mathbf{x}_i) \circ \phi(\mathbf{x}_j)$. There are several types of kernel function, and it is hard to determine the best type of kernel functions for specific data patterns [40]. However, in practice, the Gaussian radial basis functions (RBF) with a width of σ: $K(\mathbf{x}_i, \mathbf{x}_j) = \exp\left(-0.5\|\mathbf{x}_i - \mathbf{x}_j\|^2 / \sigma^2\right)$ is not only easier to implement, but also capable of nonlinearly mapping the training data into an infinite dimensional space. Therefore, the Gaussian RBF kernel function is employed in this study.

It is well known that good determination of the three parameters (including hyperparameters, C, ε, and the kernel parameter, σ) in an SVR model will seriously affect its forecasting accuracy. Thus, to look for an efficient approach to simultaneously determine well the parameter combination is becoming an important research issue. As mentioned above, inspired by Hong's series of efforts in hybridizing chaotic sequences with optimization algorithms for parameter determination to overcome the most embedded drawback of evolutionary algorithms—the premature convergence problem—this paper will continue exploring any solutions (such as empowering each particle with quantum behaviors) to overcome the embedded drawbacks of PSO, namely the QPSO algorithm, and the superiority of hybrid chaotic mapping function with the QPSO algorithms. Thus, the chaotic QPSO (CQPSO) algorithm is hybridized with an SVR model, named the SVRCQPSO model, to optimize the parameter selection to achieve more satisfactory forecasting accuracy.

2.2. Chaotic Quantum Particle Swarm Optimization Algorithm

2.2.1. Quantum Particle Swarm Optimization Algorithm

In the classical PSO algorithm, a particle's action can be addressed completely by its position and velocity which determine the trajectory of the particle, *i.e.*, any particles move along a deterministic trajectory in the search space by following Newtonian mechanics [34]. In the meanwhile, this situation also limits the possibility that the PSO algorithm could look for global optima and leads it to be trapped into local optima, *i.e.*, premature convergence. To overcome this embedded drawback of the PSO algorithm, to solve the limitation of the deterministic particle trajectory, lots of efforts in the physics literature are focused on empowering each particle trajectory with stochasticity, *i.e.*, empowering each particle's movement with quantum mechanics.

Based on Heisenberg's uncertainty principle [41], under quantum conditions, the position (x) and velocity (v) of a particle cannot be determined simultaneously, therefore, in the quantum search space, the probability of finding a particle at a particular position should be, via a "collapsing" process, mapped into its certain position in the solution space. Eventually, by employing the Monte Carlo method, the position of a particle can be updated using Equation (6):

$$x(t+1) = p(t) \pm \frac{1}{2}L(t)\ln\left(\frac{1}{u(t)}\right) \tag{6}$$

where $u(t)$ is a uniform random number distributed in [0, 1]; $p(t)$ is the particle's local attractor, and it is defined as Equation (7):

$$p(t) = \beta p_{id}(t) + (1 - \beta)p_{gd}(t) \tag{7}$$

where β is also a random number uniformly distributed in [0, 1]; $p_{id}(t)$ and $p_{gd}(t)$ are the ith *pbest* particle and the *gbest* particle in the dth dimension, respectively. $L(t)$ is the length of the potential field [35], and is given by Equation (8):

$$L(t) = 2\gamma |p(t) - x(t)| \tag{8}$$

where parameter γ is the so-called the creativity coefficient or contraction expansion coefficient, and is used to control the convergence speed of the particle. QPSO algorithm can obtain good results by linear decreasing value of γ from 1.0 to 0.5, as shown in Equation (9) [42]:

$$\gamma = (1 - 0.5) \times (Iter_{max} - t) / Iter_{max} + 0.5 \tag{9}$$

where $Iter_{max}$ is the maximum of iteration numbers, in this paper, it is set as 10,000.

Considering that the critical position of $L(t)$ will seriously influence the convergence rate and the performance of the QPSO algorithm, thus, we define the mean best position (*mbest*) as the center of *pbest* position of the swarm, shown in Equation (10):

$$mbest(t) = (mbest_1(t), \ mbest_2(t), \ \ldots, \ mbest_D(t)$$
$$= \left(\frac{1}{S}\sum_{i=1}^{S} p_{i1}(t), \ \frac{1}{S}\sum_{i=1}^{S} p_{i2}(t), \ \ldots, \ \frac{1}{S}\sum_{i=1}^{S} p_{iD}(t) \right) \tag{10}$$

where S is the size of population, D is the number of dimensions, $p_{ij}(t)$ is the *pbest* position of each particle in the jth dimension.

Then, we use Equation (10) to replace the $p(t)$ in Equation (8), thus, the new evaluation equation of $L(t)$ is Equation (11):

$$L(t) = 2\gamma |mbest(t) - x(t)| \tag{11}$$

Finally, by substituting Equations (7) and (11) into Equation (6), the particle's position is updated by Equation (12):

$$x(t + 1) = \beta p_{id}(t) + (1 - \beta)p_{gd}(t) \pm \gamma |mbest(t) - x(t)| \ln\left(\frac{1}{u(t)}\right) \tag{12}$$

2.2.2. Chaotic Mapping Function for QPSO Algorithm

As mentioned that chaotic variable can be adopted by applying chaotic phenomenon in keeping the diversities among particles to prevent the PSO algorithm from being trapped into a local optima, *i.e.*, premature convergence. Therefore, the CQPSO algorithm is based on the QPSO algorithm by employing chaotic strategy while premature convergence appears during the iterative searching processes, else, the QPSO algorithm is still implemented as illustrated in Section 2.2.1.

On the other hand, for strengthening the effect of chaotic characteristics, lots of studies mostly apply the logistic mapping function as chaotic sequence generator. The biggest disadvantage of the logistic mapping function is that it distributes at both ends and less in the middle. On the contrary, the Cat mapping function has better chaotic distribution characteristic, thus, its application in chaos disturbance of the PSO algorithm can better strengthen the swarm diversity [43]. Therefore, this paper will employ the Cat mapping function as chaotic sequence generator.

The classical Cat mapping function is the two-dimensional Cat mapping function [44], shown as Equation (13):

$$\begin{cases} x_{n+1} = (x_n + y_n) \bmod 1 \\ y_{n+1} = (x_n + 2y_n) \bmod 1 \end{cases} \tag{13}$$

where $x \bmod 1 = x - [x]$, mod, the so-called modulo operation, is used for the fractional parts of a real number x by subtracting an appropriate integer.

2.2.3. Implementation Steps of CQPSO Algorithm

The procedure of hybrid CQPSO algorithm with an SVR model is illustrated as follows and the corresponding flowchart is shown as Figure 1.

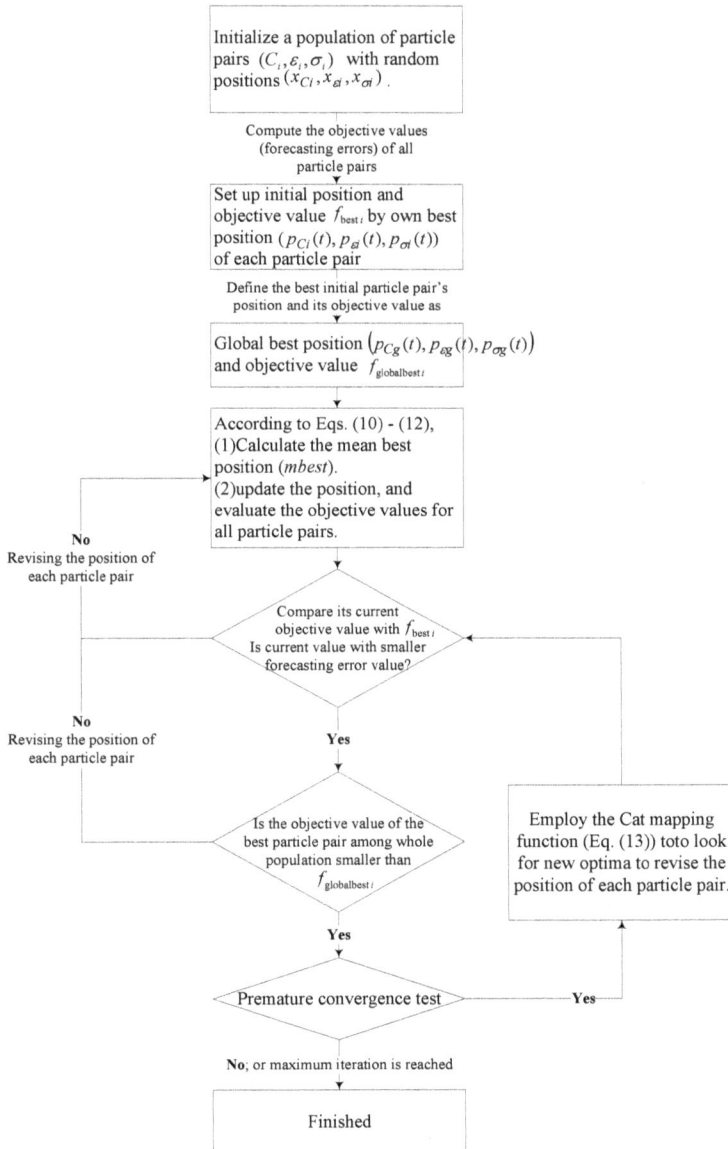

Figure 1. Quantum particle swarm optimization flowchart.

Step 1: Initialization.

Initialize a defined population of particle pairs $(C_i, \varepsilon_i, \sigma_i)$ with random positions $(x_{Ci}, x_{\varepsilon i}, x_{\sigma i})$, where each particle contains n variables.

Step 2: Objective Values.

Compute the objective values (forecasting errors) of all particle pairs. Let the particle's own best position be $p_{id}(t) = (p_{Ci}(t), p_{\varepsilon i}(t), p_{\sigma i}(t))$ of each particle pair and its objective value $f_{\text{best } i}$ equal its initial position and objective value. Let the global best position be $p_{gd}(t) = (p_{Cg}(t), p_{\varepsilon g}(t), p_{\sigma g}(t))$ and its objective value $f_{\text{globalbest } i}$ equal to the best initial particle pair's position and its objective value.

Step 3: Calculate Objective Values.

Employ Equation (10) to calculate the mean best position (*m*best), the center of *p*best position of the three particle pairs, then, use Equations (11) and (12) to update the position for each particle pair, and calculate the objective values for all particle pairs.

Step 4: Update.

For each particle pair, compare its current objective value with $f_{\text{best } i}$. If current value is better (with smaller forecasting accuracy index value), then, update $(p_{Ci}(t), p_{\varepsilon i}(t), p_{\sigma i}(t))$ and its objective value with the current position and objective value.

Step 5: Determine the Best Position and Objective.

Determine the best particle pair of whole population based on the best objective value. If the objective value is smaller than $f_{\text{globalbest } i}$, then update $(p_{Cg}(t), p_{\varepsilon g}(t), p_{\sigma g}(t))$, and, use Equation (7) to update the particle pair's local attractor. Finally, update its objective value with the current best particle pair's position.

Step 6: Premature Convergence Test.

Calculate the mean square error (MSE), shown as Equation (14), to evaluate the premature convergence status, set the expected criteria, δ:

$$\text{MSE} = \frac{1}{S} \sum_{i=1}^{S} \left(\frac{f_i - f_{avg}}{f} \right)^2 \tag{14}$$

where f_i is the current objective value of the current particles; f_{avg} is average objective value of the current swarm; f can be obtained by Equation (15):

$$f = \max \left\{ 1, \max_{\forall i \in S} \left\{ |f_i - f_{avg}| \right\} \right\} \tag{15}$$

If the value of MSE is less than δ, it can be seen that premature convergence appears. Thus, the Cat mapping function, Equation (13), is then employed to look for new optima, and set the new optimal value as the optimal solution of the current particles.

Step 7: Stop Criteria.

If a stopping threshold (forecasting accuracy) is reached, then $(P_{Cg}, P_{\varepsilon g}, P_{\sigma g})$ and its $f_{\text{globalbest } i}$ would be determined; otherwise go back to Step 3.

In this paper, the mean absolute percentage error (MAPE) as the forecasting accuracy index, shown in Equation (16), is employed for calculating the objective value to determine suitable parameters in Steps 4 and 5 of QPSO algorithm:

$$\text{MAPE} = \frac{1}{N} \sum_{i=1}^{N} \left| \frac{y_i - f_i}{y_i} \right| \times 100\% \tag{16}$$

where N is the number of forecasting periods; y_i is the actual value at period i; f_i denotes is the forecasting value at period i.

3. Numerical Examples

3.1. Data Set of Numerical Examples

3.1.1. Regional Load Data

The first numerical example applies Taiwan regional electric demand data from an existing published paper [33] to construct the proposed SVRCQPSO model, and the forecasting accuracy of the proposed model and other alternatives is compared. Therefore, in this example, the total load values in four regions of Taiwan from 1981 to 2000 (20 years) serve as experimental data. To be based on the same comparison basis, these load data are divided into three subsets, the training data set (from 1981 to 1992, *i.e.*, 12 load data), the validation data set (from 1993 to 1996, that is four load data), and the testing data set (from 1997 to 2000, *i.e.*, four load data). The forecasting accuracy is measured by Equation (16).

During the training process, the rolling-based forecasting procedure proposed by Hong [33] is employed, which divides training data into two subsets, namely fed-in (eight load data) and fed-out (four load data) respectively. The training error can be obtained in each iteration. While training error is decreasing, the three parameters determined by QPSO algorithm are employed to calculate the validation error. Then, those parameters with minimum validation error are selected as the most appropriate candidates. Notice that the testing data set is never employed while modeling. Eventually, the desired four-years forecasting loads in each region are forecasted. Along with the smallest testing MAPE value, the proposed model is the most suitable model in this example.

3.1.2. Annual Load Data

The second numerical example also uses Taiwan annual electric demand data from an existing paper [33]. The total annual electric demand values from 1945 to 2003 (59 years) serve as the experimental data. To be based on the same comparison basis, these employed load data are also divided into three data sets, the training data set (from 1945 to 1984, *i.e.*, 40 years), the validation data set (from 1985 to 1994, that is 10 years), and the testing data set (from 1995 to 2003, *i.e.*, nine years). Similarly, the forecasting accuracy is also measured by MAPE. Meanwhile, the rolling-based forecasting procedure, the structural risk minimization principle to minimize the training error, the procedure to determine parameter combination, and so on, are also implemented the same as in the first numerical example.

3.1.3. Load Data in 2014 Global Energy Forecasting Competition (GEFCOM 2014)

The third numerical example is suggested to use the historical hourly load data issued in 2014 Global Energy Forecasting Competition [45]. The total hourly load values, from 00:00 1 December 2011 to 00:00 1 January 2012 (744 h), serve as experimental data. These load data are divided into three data sets, the training data set (from 01:00 1 December 2011 to 00:00 24 December 2011, *i.e.*, 552 h load data), the validation data set (from 01:00 24 December 2011 to 00:00 18 December 2011, that is 96 h load data), and the testing data set (from 01:00 28 December 2011 to 00:00 1 January 2012, *i.e.*, 96 h load data). Similarly, the forecasting accuracy is also measured by MAPE; the rolling-based forecasting procedure, the structural risk minimization principle to minimize the training error, and the procedure to determine parameter combination are also implemented as the same as in the previous two numerical examples.

3.2. The SVRCQPSO Load Forecasting Model

3.2.1. Parameter Setting in the CQPSO Algorithm

Proper tuning of control parameters for convergence of the classical PSO algorithm is not easy, on the contrary, there is only one parameter control in the CQPSO algorithm, *i.e.*, the creativity coefficient or contraction expansion coefficient, γ, given by Equation (9). Other settings, such as the population sizes, are 20 in both examples; the total number of iterations (*Iter*$_{max}$) is both fixed as 10,000; $\sigma \in [0, 5]$, $\varepsilon \in [0, 100]$ in both examples, $C \in [0, 20000]$ in example one, $C \in [0, 3 \times 10^{10}]$ in example two; δ is both set as 0.001.

3.2.2. Three Parameter Determination of SVRQPSO and SVRCQPSO Models in Regional Load Data

For the first numerical example, the potential models with well determined parameter values by QPSO algorithm and CQPSO algorithm which have the smallest testing MAPE value will be selected as the most suitable models. The determined parameters for four regions in Taiwan are illustrated in Table 1.

Table 1. Parameters determination of SVRCQPSO and SVRQPSO models (example one).

Regions	SVRCQPSO Parameters			MAPE of Testing (%)
	σ	C	ε	
Northern	10.0000	0.9000×10^{10}	0.7200	1.1070
Central	10.0000	1.8000×10^{10}	0.4800	1.2840
Southern	4.0000	0.8000×10^{10}	0.2500	1.1840
Eastern	3.0000	1.2000×10^{10}	0.3400	1.5940

Regions	SVRQPSO Parameters			MAPE of Testing (%)
	σ	C	ε	
Northern	8.0000	1.4000×10^{10}	0.6500	1.3370
Central	8.0000	0.8000×10^{10}	0.4300	1.6890
Southern	4.0000	0.6000×10^{10}	0.6500	1.3590
Eastern	12.0000	1.0000×10^{10}	0.5600	1.9830

Meanwhile, based on the same forecasting duration in each region, Table 2 shows the MAPE values and forecasting results of various forecasting models in each region, including SVRCQPSO (hybridizing chaotic function, quantum mechanics, and PSO with SVR), SVRQPSO (hybridizing quantum mechanics and PSO with SVR), SVMG (hybridizing genetic algorithm with SVM), and RSVMG (hybridizing recurrent mechanism and genetic algorithm with SVM) models. In Table 2, the SVRQPSO model has almost outperformed SVRPSO models that hybridize classical PSO algorithm with an SVR model. It also demonstrates that empowering the particles to have quantum behaviors, *i.e.*, applying quantum mechanics in the PSO algorithm, is a feasible approach to improve the solution, to improve the forecasting accuracy while the PSO algorithm is hybridized with an SVR model. In addition, the SVRCQPSO model eventually achieves a smaller MAPE value than other alternative models, except the RSVMG model in the northern region. It also illustrates that the Cat mapping function has done a good job of looking for more satisfactory solutions while suffering from the premature convergence problem during the QPSO algorithm processing. Once again, it also obviously illustrates the performance of the chaotic mapping function in overcoming the premature convergence problem. For example, in the northern region, we had done our best by using the QPSO algorithm, we could only to look for the solution, $(\sigma, C, \varepsilon) = (8.0000, 1.4000 \times 10^{10}, 0.6500)$, with forecasting error, 1.3370%, as mentioned above that it is superior to classical PSO algorithm. However, the solution still could be improved by the CQPSO algorithm to $(\sigma, C, \varepsilon) = (10.0000, 0.9000 \times 10^{10}, 0.7200)$ with more accurate forecasting performance, 1.1070%. Similarly, for other regions, the solutions of the QPSO algorithm with forecasting errors, 1.6890% (the central region), 1.3590% (the southern region) and 1.9830% (the eastern region), all could be further searched for more accurate forecasting performance by applying the Cat mapping function, *i.e.*, the CQPSO algorithm, to receive more satisfactory results, such as 1.2840% (the central region), 1.1840% (the southern region), and 1.5940% (the eastern region), respectively.

Table 2. Forecasting results of SVRCQPSO, SVRQPSO, and other models (example one) (unit: 10^6 MWh).

| Year | Northern Region | | | | | | |
	Actual	SVRCQPSO	SVRQPSO	SVRCPSO	SVRPSO	SVMG	RSVMG
1997	11,222	11,339	11,046	11,232	11,245	11,213	11,252
1998	11,642	11,779	11,787	11,628	11,621	11,747	11,644
1999	11,981	11,832	12,144	12,016	12,023	12,173	12,219
2000	12,924	12,798	12,772	12,306	12,306	12,543	12,826
MAPE (%)	-	**1.1070**	1.3370	1.3187	1.3786	1.3891	0.7498
Year	Central Region						
	Actual	SVRCQPSO	SVRQPSO	SVRCPSO	SVRPSO	SVMG	RSVMG
1997	5061	4987	5140	5066	5085	5060	5065
1998	5246	5317	5342	5168	5141	5203	5231
1999	5233	5172	5130	5232	5236	5230	5385
2000	5633	5569	5554	5313	5343	5297	5522
MAPE (%)	-	**1.2840**	1.6890	1.8100	1.9173	1.8146	1.3026
Year	Southern Region						
	Actual	SVRCQPSO	SVRQPSO	SVRCPSO	SVRPSO	SVMG	RSVMG
1997	6336	6262	6265	6297	6272	6265	6200
1998	6318	6401	6418	6311	6314	6389	6156
1999	6259	6179	6178	6324	6327	6346	6261
2000	6804	6738	6901	6516	6519	6513	6661
MAPE (%)	-	**1.1840**	1.3590	1.4937	1.5899	2.0243	1.7530
Year	Eastern Region						
	Actual	SVRCQPSO	SVRQPSO	SVRCPSO	SVRPSO	SVMG	RSVMG
1997	358	353	350	370	367	358	367
1998	397	404	390	376	374	373	381
1999	401	394	410	411	409	397	401
2000	420	414	413	418	415	408	416
MAPE (%)	-	**1.5940**	1.9830	2.1860	2.3094	2.6475	1.8955

Furthermore, to ensure the significant improvement in forecasting accuracy for the proposed SVRQPSO and SVRCQPSO models, as Diebold and Mariano [46] recommend, a suitable statistical test, namely the Wilcoxon signed-rank test, is then implemented. The test can be implemented at two different significance levels, *i.e.*, $\alpha = 0.025$ and $\alpha = 0.05$, by one-tail-tests. The test results are shown in Table 3, which indicates that the SVRCQPSO model only achives significantly better performance than other alternatives in the northern and eastern regions in terms of MAPE. It also implies that in these two regions, the load tendency is approaching a mature status, *i.e.*, in northern Taiwan, it is highly commercial and residential electricity usage type; in eastern Taiwan, the highly concentrated natural resources only reflects its low electricity usage type. In both regions, the electricity load tendency and trend no doubt could be easily captured by the proposed SVRCQPSO model, thus, the proposed SVRCQPSO model can significantly outperform other alternatives.

Table 3. Wilcoxon signed-rank test (example one).

Compared Models	Wilcoxon Signed-Rank Test							
	$\alpha = 0.025$; W = 0				$\alpha = 0.05$; W = 0			
	Northern Region	Central Region	Southern Region	Eastern Region	Northern Region	Central Region	Southern Region	Eastern Region
SVRCQPSO *vs.* SVMG	0 [a]	1	1	0 [a]	0 [a]	1	1	0 [a]
SVRCQPSO *vs.* RSVMG	1	1	0 [a]	0 [a]	1	1	0 [a]	0 [a]
SVRCQPSO *vs.* SVRPSO	0 [a]	1	1	0 [a]	0 [a]	1	1	0 [a]
SVRCQPSO *vs.* SVRCPSO	0 [a]	1	1	0 [a]	0 [a]	1	1	0 [a]
SVRCQPSO *vs.* SVRQPSO	1	1	0 [a]	0 [a]	1	1	0 [a]	0 [a]

[a] denotes that the SVRCQPSO model significantly outperforms other alternative models.

On the other hand, in the central and southern regions, the SVRCQPSO model almost could not achieve significant accuracy improvements compared to the other models. It also reflects the facts that these two regions in Taiwan are both high-density population centers, the electricity usage types would be very flexible almost along with population immigration or emigration, thus, although the proposed SVRCQPSO model captures the data tendencies this time, however, it could not guarantee it will also achieve highly accurate forecasting performance when new data is obtained. Therefore, this is also the next research topic.

3.2.3. Three Parameters Determination of SVRQPSO and SVRCQPSO Models in Annual Load Data

For the second numerical example, the processing steps are similar to the example one. The parameters in an SVR model will also be determined by the proposed QPSO algorithm and CQPSO algorithm. Then, the selected models would be with the smallest testing MAPE values. The determined parameters for annual loads in Taiwan (example two) are illustrated in Table 4. For benchmarking comparison with other algorithms, Table 4 lists all results in relevant papers with SVR-based modeling, such as the Pai and Hong [47] proposed SVMSA model by employing SA algorithm and the Hong [33] proposed SVRCPSO and SVRPSO models by using the CPSO algorithm and PSO algorithm, respectively.

Table 4. Parameter determination of SVRCQPSO and SVRQPSO models (example two).

Optimization Algorithms	Parameters			MAPE of Testing (%)
	σ	C	ε	
SA algorithm [46]	0.2707	2.8414×10^{11}	39.127	1.7602
PSO algorithm [33]	0.2293	1.7557×10^{11}	10.175	3.1429
CPSO algorithm [33]	0.2380	2.3365×10^{11}	39.296	1.6134
QPSO algorithm	12.0000	0.8000×10^{11}	0.380	1.3460
CQPSO algorithm	10.0000	1.5000×10^{11}	0.560	1.1850

Figure 2 illustrates the real values and forecasting values of different models, including the hybridizing simulated annealing algorithm with SVM (SVMSA), SVRPSO, SVRCPSO, SVRQPSO, and SVRCQPSO models. In Table 4, similarly, the SVRQPSO model is superior to SVRPSO models that hybridize a classical PSO algorithm with an SVR model. Once again, it also demonstrates that applying quantum mechanics in the PSO algorithm is a feasible approach to improve the forecasting accuracy of any SVR-based forecasting model. In addition, the SVRCQPSO model eventually achieves the smallest MAPE value than other alternative models. Of course, the Cat mapping function provides its excellent improvement in overcoming the premature convergence problem. It can be clearly to see that based on the QPSO algorithm, we could only look for the solution, $(\sigma, C, \varepsilon) = (12.0000, 0.8000 \times 10^{11}, 0.380)$, with a 1.3460% forecasting error, although it is superior to the classical PSO algorithm. Then, the Cat mapping function is excellent to shift the solution of the QPSO algorithm to another better solution, $(\sigma, C, \varepsilon) = (10.0000, 1.5000 \times 10^{11}, 0.560)$ with a forecasting error of 1.1850%.

To verify the significance of the proposed SVRCQPSO model in this annual load forecasting example, similarly, the Wilcoxon signed-rank test is also taken into account. The test results are shown in Table 5, which indicate that the SVRCQPSO model has completely achieved a more significant performance than other alternatives in terms of MAPE, *i.e.*, the annual load tendency in Taiwan reflects an increasing trend due to the strong annual economic growth. The electricity load tendency and trend no doubt could be easily captured by the proposed SVRCQPSO model; therefore, the proposed SVRCQPSO model can significantly outperform other alternatives.

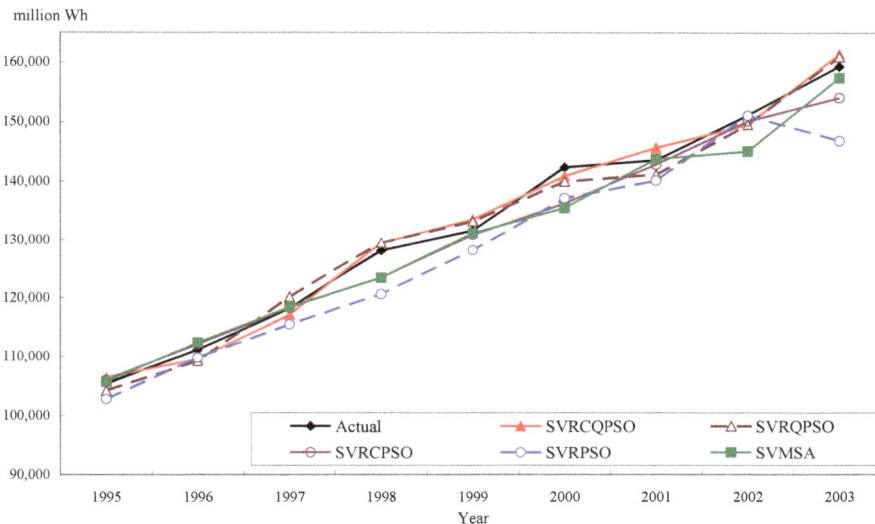

Figure 2. Actual values and forecasting values of SVRCQPSO, SVRQPSO, and other models (example two).

Table 5. Wilcoxon signed-rank test (example two).

Compared Models	Wilcoxon Signed-Rank Test	
	$\alpha = 0.025; W = 5$	$\alpha = 0.05; W = 8$
SVRCQPSO *vs.* SVMSA	2 [a]	2 [a]
SVRCQPSO *vs.* SVRPSO	3 [a]	3 [a]
SVRCQPSO *vs.* SVRCPSO	2 [a]	2 [a]
SVRCQPSO *vs.* SVRQPSO	2 [a]	2 [a]

[a] denotes that the SVRCQPSO model significantly outperforms other alternative models.

3.2.4. Three Parameter Determination of SVRQPSO and SVRCQPSO Models in GEFCOM 2014

For the third numerical example, the processing steps are to be conducted similarly. The determined parameters in an SVR model by the proposed QPSO algorithm and CQPSO algorithm will have the smallest MAPE values in the test data set. The determined parameters for GEFCOM 2014 (example three) are illustrated in Table 6. In addition, the parameters determined by other famous algorithms, such as GA, CGA, PSO, CPSO algorithms, are also listed in Table 6. Because GEFCOM 2014 load data is a completely new case for the author, to correctly assess the improvements of the proposed models, a naïve model is introduced, which is appropriately to be a random search of the hyper-parameters. Therefore, the randomly determined parameters are also illustrated in Table 6.

Table 6. Parameters determination of SVRCQPSO and SVRQPSO models (example three).

Optimization Algorithms	Parameters			MAPE of Testing (%)
	σ	C	ε	
Naïve	23.000	43.000	0.6700	3.2200
CGA	19.000	28.000	0.2700	2.9100
PSO algorithm	7.000	34.000	0.9400	3.1500
CPSO algorithm	22.000	19.000	0.6900	2.8600
QPSO algorithm	9.000	42.000	0.1800	1.9600
CQPSO algorithm	19.000	35.000	0.8200	1.2900

For the forecasting performance comparison, the author also considers two famous forecasting models, the ARIMA(0, 1, 1) model, and the back propagation neural networks (BPNN) model to conduct benchmark comparisons. Figure 3 illustrates the real values and forecasting results, including the ARIMA, BPNN, Naïve, SVRCGA, SVRPSO, SVRCPSO, SVRQPSO, and SVRCQPSO models. In Figure 3, it also indicates that the SVRQPSO model achives more accurate forecasting performance than the SVRPSO and SVRCPSO models that hybridize classical PSO algorithms or chaotic sequences with an SVR model. It also illustrates the application of quantum mechanics in the PSO algorithm is a potential approach to improve the performance issues of any SVR-based model. In addition, the SVRCQPSO model eventually achieves a smaller MAPE value than the SVRQPSO model.

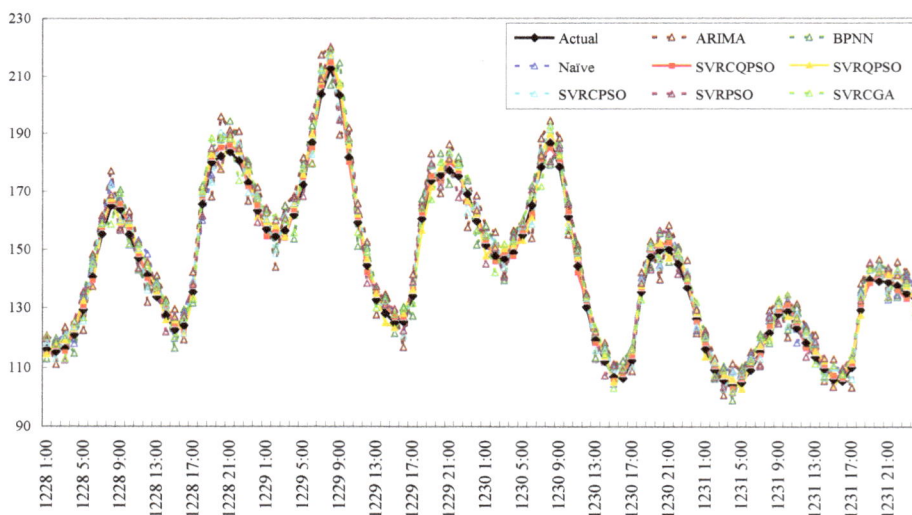

Figure 3. Actual values and forecasting values of SVRCQPSO, SVRQPSO, and other models (example three).

Finally, the results of Wilcoxon signed-rank test are presented in Table 7, which indicates that the proposed SVRCQPSO model achieves superior significance in terms of MAPE, *i.e.*, the hourly electric load reflects a cyclic trend which is captured exactly by the proposed SVRCQPSO model; therefore, the proposed SVRCQPSO model can significantly outperform other alternatives.

Table 7. Wilcoxon signed-rank test (example three).

Compared Models	Wilcoxon Signed-Rank Test	
	α = 0.025; W = 2,328	α = 0.05; W = 2,328
SVRCQPSO *vs.* ARIMA	1612 [a]	1612 [a]
SVRCQPSO *vs.* BPNN	1715 [a]	1715 [a]
SVRCQPSO *vs.* Naïve	1650 [a]	1650 [a]
SVRCQPSO *vs.* SVRPSO	1713 [a]	1713 [a]
SVRCQPSO *vs.* SVRCPSO	1654.5 [a]	1654.5 [a]
SVRCQPSO *vs.* SVRQPSO	1700 [a]	1700 [a]
SVRCQPSO *vs.* SVRCGA	1767 [a]	1767 [a]

[a] denotes that the SVRCQPSO model significantly outperforms other alternative models.

4. Conclusions

This paper presents an SVR model hybridized with the chaotic Cat mapping function and quantum particle swarm optimization algorithm (CQPSO) for electric demand forecasting. The experimental results demonstrate that the proposed model obtains the best forecasting performance among other SVR-based forecasting models in the literature, even though overall the forecasting superiority does not meet the significance test. This paper applies quantum mechanics to empower particles to have quantum behaviors to improve the premature convergence of the PSO algorithm and then, improve the forecasting accuracy. Chaotic Cat mapping is also employed to help with unexpected trapping into local optima while the QPSO algorithm is working in its searching process. This paper also illustrates the good feasibility of hybridizing quantum mechanics to expand the search space which is usually limited by Newtonian dynamics. In future research, as mentioned in Section 3.2.2, how to enhance the power of the QPSO algorithm to capture the tendency changes of electricity load data along with population immigration or emigration to guarantee the SVRCQPSO model achieves highly accurate forecasting performance will be studied.

Conflicts of Interest: The author declares no conflict of interest.

References

1. Xiao, L.; Wang, J.; Hou, R.; Wu, J. A combined model based on data pre-analysis and weight coefficients optimization for electrical load forecasting. *Energy* **2015**, *82*, 524–549. [CrossRef]
2. Bunn, D.W.; Farmer, E.D. Comparative models for electrical load forecasting. *Int. J. Forecast.* **1986**, *2*, 241–242.
3. Zhao, W.; Wang, J.; Lu, H. Combining forecasts of electricity consumption in China with time-varying weights updated by a high-order Markov chain model. *Omega* **2014**, *45*, 80–91. [CrossRef]
4. Lee, C.M.; Ko, C.N. Short-term load forecasting using lifting scheme and ARIMA models. *Expert Syst. Appl.* **2011**, *38*, 5902–5911. [CrossRef]
5. Taylor, J.W.; Snyder, R.D. Forecasting intraday time series with multiple seasonal cycles using parsimonious seasonal exponential smoothing. *Omega* **2012**, *40*, 748–757. [CrossRef]
6. Hippert, H.S.; Taylor, J.W. An evaluation of Bayesian techniques for controlling model complexity and selecting inputs in a neural network for short-term load forecasting. *Neural Netw.* **2010**, *23*, 386–395. [CrossRef] [PubMed]
7. Al-Hamadi, H.M.; Soliman, S.A. Short-term electric load forecasting based on Kalman filtering algorithm with moving window weather and load model. *Electr. Power Syst. Res.* **2004**, *68*, 47–59. [CrossRef]
8. Zheng, T.; Girgis, A.A.; Makram, E.B. A hybrid wavelet-Kalman filter method for load forecasting. *Electr. Power Syst. Res.* **2000**, *54*, 11–17. [CrossRef]
9. Dudek, G. Pattern-based local linear regression models for short-term load forecasting. *Electr. Power Syst. Res.* **2016**, *130*, 139–147. [CrossRef]
10. Li, H.Z.; Guo, S.; Li, C.J.; Sun, J.Q. A hybrid annual power load forecasting model based on generalized regression neural network with fruit fly optimization algorithm. *Knowl. Based Syst.* **2013**, *37*, 378–387. [CrossRef]

11. Hong, W.C. Electric load forecasting by seasonal recurrent SVR (support vector regression) with chaotic artificial bee colony algorithm. *Energy* **2011**, *36*, 5568–5578. [CrossRef]

12. Lou, C.W.; Dong, M.C. A novel random fuzzy neural networks for tackling uncertainties of electric load forecasting. *Int. J. Electr. Power Energy Syst.* **2015**, *73*, 34–44. [CrossRef]

13. Niu, D.X.; Shi, H.; Wu, D.D. Short-term load forecasting using bayesian neural networks learned by Hybrid Monte Carlo algorithm. *Appl. Soft Comput.* **2012**, *12*, 1822–1827. [CrossRef]

14. Hooshmand, R.A.; Amooshahi, H.; Parastegari, M. A hybrid intelligent algorithm based short-term load forecasting approach. *Int. J. Electr. Power Energy Syst.* **2013**, *45*, 313–324. [CrossRef]

15. Li, S.; Wang, P.; Goel, L. Short-term load forecasting by wavelet transform and evolutionary extreme learning machine. *Electr. Power Syst. Res.* **2015**, *122*, 96–103. [CrossRef]

16. Hanmandlu, M.; Chauhan, B.K. Load forecasting using hybrid models. *IEEE Trans. Power Syst.* **2011**, *26*, 20–29. [CrossRef]

17. Mahmoud, T.S.; Habibi, D.; Hassan, M.Y.; Bass, O. Modelling self-optimised short term load forecasting for medium voltage loads using tunning fuzzy systems and artificial neural networks. *Energy Convers Manag.* **2015**, *106*, 1396–1408. [CrossRef]

18. Suykens, J.A.K.; Vandewalle, J.; De Moor, B. Optimal control by least squares support vector machines. *Neural Netw.* **2001**, *14*, 23–35. [CrossRef]

19. Sankar, R.; Sapankevych, N.I. Time series prediction using support vector machines: A survey. *IEEE Commun. Mag.* **2009**, *4*, 24–38.

20. Hahn, H.; Meyer-Nieberg, S.; Pickl, S. Electric load forecasting methods: Tools for decision making. *Eur. J. Oper. Res.* **2009**, *199*, 902–907. [CrossRef]

21. Drucker, H.; Burges, C.C.J.; Kaufman, L.; Smola, A.; Vapnik, V. Support vector regression machines. *Adv. Neural Inform. Process. Syst.* **1997**, *9*, 155–161.

22. Chen, Y.H.; Hong, W.C.; Shen, W.; Huang, N.N. Electric load forecasting based on LSSVM with fuzzy time series and global harmony search algorithm. *Energies* **2016**, *9*, 70. [CrossRef]

23. Fan, G.; Peng, L.-L.; Hong, W.-C.; Sun, F. Electric load forecasting by the SVR model with differential empirical mode decomposition and auto regression. *Neurocomputing* **2016**, *173*, 958–970. [CrossRef]

24. Geng, J.; Huang, M.L.; Li, M.W.; Hong, W.C. Hybridization of seasonal chaotic cloud simulated annealing algorithm in a SVR-based load forecasting model. *Neurocomputing* **2015**, *151*, 1362–1373. [CrossRef]

25. Ju, F.Y.; Hong, W.C. Application of seasonal SVR with chaotic gravitational search algorithm in electricity forecasting. *Appl. Math. Model.* **2013**, *37*, 9643–9651. [CrossRef]

26. Fan, G.; Wang, H.; Qing, S.; Hong, W.C.; Li, H.J. Support vector regression model based on empirical mode decomposition and auto regression for electric load forecasting. *Energies* **2013**, *6*, 1887–1901. [CrossRef]

27. Hong, W.C.; Dong, Y.; Zhang, W.Y.; Chen, L.Y.; Panigrahi, B.K. Cyclic electric load forecasting by seasonal SVR with chaotic genetic algorithm. *Int. J. Electr. Power Energy Syst.* **2013**, *44*, 604–614. [CrossRef]

28. Zhang, W.Y.; Hong, W.C.; Dong, Y.; Tsai, G.; Sung, J.T.; Fan, G. Application of SVR with chaotic GASA algorithm in cyclic electric load forecasting. *Energy* **2012**, *45*, 850–858. [CrossRef]

29. Hong, W.C.; Dong, Y.; Lai, C.Y.; Chen, L.Y.; Wei, S.Y. SVR with hybrid chaotic immune algorithm for seasonal load demand forecasting. *Energies* **2011**, *4*, 960–977. [CrossRef]

30. Hong, W.C. Application of chaotic ant swarm optimization in electric load forecasting. *Energy Policy* **2010**, *38*, 5830–5839. [CrossRef]

31. Hong, W.C. Hybrid evolutionary algorithms in a SVR-based electric load forecasting model. *Int. J. Electr. Power Energy Syst.* **2009**, *31*, 409–417. [CrossRef]

32. Hong, W.C. Electric load forecasting by support vector model. *Appl. Math. Model.* **2009**, *33*, 2444–2454. [CrossRef]

33. Hong, W.C. Chaotic particle swarm optimization algorithm in a support vector regression electric load forecasting model. *Energy Convers Manag.* **2009**, *50*, 105–117. [CrossRef]

34. Kennedy, J.; Eberhart, R.C. Particle swarm optimization. In Proceedings of the IEEE International Conference on Neural Networks, Piscataway, NJ, USA, 27 November–1 December 1995; pp. 1942–1948.

35. Sun, J.; Feng, B.; Xu, W.B. Particle swarm optimization with particles having quantum behavior. In Proceedings of the IEEE Proceedings of Congress on Evolutionary Computation, Piscataway, NJ, USA, 19–23 June 2004; pp. 325–331.

36. Davoodi, E.; Haque, M.T.; Zadeh, S.G. A hybrid improved quantum-behaved particle swarm optimization—Simplex method (IQPSOS) to solve power system load flow problems. *Appl. Soft Comput.* **2014**, *21*, 171–179. [CrossRef]

37. Kamberaj, H. Q-Gaussian swarm quantum particle intelligence on predicting global minimum of potential energy function. *Appl. Math. Comput.* **2014**, *229*, 94–106. [CrossRef]

38. Li, Y.; Jiao, L.; Shang, R.; Stolkin, R. Dynamic-context cooperative quantum-behaved particle swarm optimization based on multilevel thresholding applied to medical image segmentation. *Inform. Sci.* **2015**, *294*, 408–422. [CrossRef]

39. Coelho, L.D.S. A quantum particle swarm optimizer with chaotic mutation operator. *Chaos Solitons Fractals* **2008**, *37*, 1409–1418. [CrossRef]

40. Amari, S.; Wu, S. Improving support vector machine classifiers by modifying kernel functions. *Neural Netw.* **1999**, *12*, 783–789. [CrossRef]

41. Heisenberg, W. Über den anschaulichen Inhalt der quantentheoretischen Kinematik und Mechanik. *Z. Phys.* **1927**, *43*, 172–198. (In German) [CrossRef]

42. Liu, F.; Zhou, Z. An improved QPSO algorithm and its application in the high-dimensional complex problems. *Chemom. Intell. Lab. Syst.* **2014**, *132*, 82–90. [CrossRef]

43. Li, M.; Hong, W.C.; Kang, H. Urban traffic flow forecasting using Gauss-SVR with cat mapping, cloud model and PSO hybrid algorithm. *Neurocomputing* **2013**, *99*, 230–240. [CrossRef]

44. Chen, G.; Mao, Y.; Chui, C.K. Asymmetric image encryption scheme based on 3D chaotic cat maps. *Chaos Solitons Fractals* **2004**, *21*, 749–761. [CrossRef]

45. 2014 Global Energy Forecasting Competition. Available online: http://www.drhongtao.com/gefcom/ (accessed on 28 May 2016).

46. Diebold, F.X.; Mariano, R.S. Comparing predictive accuracy. *J. Bus. Econ. Stat.* **1995**, *13*, 134–144.

47. Pai, P.F.; Hong, W.C. Support vector machines with simulated annealing algorithms in electricity load forecasting. *Energy Convers Manag.* **2005**, *46*, 2669–2688. [CrossRef]

energies

MDPI

Article

Application of Hybrid Quantum Tabu Search with Support Vector Regression (SVR) for Load Forecasting

Cheng-Wen Lee [1] and Bing-Yi Lin [2,*]

[1] Department of International Business, Chung Yuan Christian University/200 Chung Pei Rd.,
    Chungli District, Taoyuan City 32023, Taiwan; chengwen@cycu.edu.tw
[2] Ph.D. Program in Business, College of Business, Chung Yuan Christian University/200 Chung Pei Rd.,
    Chungli District, Taoyuan City 32023, Taiwan
* Correspondence: g10304612@cycu.edu.tw; Tel.: +886-2-2596-8821; Fax: +886-2-2552-8844

Academic Editor: Wei-Chiang Hong
Received: 22 July 2016; Accepted: 10 October 2016; Published: 26 October 2016

Abstract: Hybridizing chaotic evolutionary algorithms with support vector regression (SVR) to improve forecasting accuracy is a hot topic in electricity load forecasting. Trapping at local optima and premature convergence are critical shortcomings of the tabu search (TS) algorithm. This paper investigates potential improvements of the TS algorithm by applying quantum computing mechanics to enhance the search information sharing mechanism (tabu memory) to improve the forecasting accuracy. This article presents an SVR-based load forecasting model that integrates quantum behaviors and the TS algorithm with the support vector regression model (namely SVRQTS) to obtain a more satisfactory forecasting accuracy. Numerical examples demonstrate that the proposed model outperforms the alternatives.

Keywords: support vector regression (SVR); quantum tabu search (QTS) algorithm; quantum computing mechanics; electric load forecasting

1. Introduction

A booming economy is dramatically increasing electric loads in every industry and those associated with people's daily lives. Meeting the demand of all has become an important goal of electricity providers. However, as mentioned by Bunn and Farmer [1], a 1% increase in the error in an electricity demand forecast corresponds to a £10 million increase in operating costs. Therefore, decision-makers seek accurate load forecasting to set effective energy policies, such as those concerning new power plants and investment in facilities [2]. Importing or exporting electricity in energy-limited developing economies, such as that of Taiwan, is almost impossible [3,4]. Unfortunately, electric load data have various characteristics, including nonlinearity and chaos. Moreover, many exogenous factors interact with each other, affecting forecasting, such as economic activities, weather conditions, population, industrial production, and others. These effects increase the difficulty of load forecasting [5].

In the last few decades, models for improving the accuracy of load forecasting have included the well-known Box–Jenkins' ARIMA model [6], exponential smoothing model [7], Kalman filtering/linear quadratic estimation model [8–10], the Bayesian estimation model [11–13], and regression models [14–16]. However, most of these models are theoretically based on assumed linear relationships between historical data and exogenous variables and so cannot effectively capture the complex nonlinear characteristics of load series, or easily provide highly accurate load forecasting.

Since the 1980s, to improve the accuracy of load forecasting, many artificial intelligent (AI) approaches have been used and been combined to develop powerful forecasting methods, such as artificial neural networks (ANNs) [17–21], expert system-based methods [22–24], and fuzzy inference approaches [25–28]. Recently, these AI approaches have been hybridized with each other to provide more accurate forecasting results [29–33], with the aforementioned linear models [34], or with evolutionary algorithms [35,36]. However, the shortcomings of these AI approaches include the need to determine the structural parameters [37,38], the time required for knowledge acquisition [39], and a lack of correct and consistent heuristic rules to generate a complete domain knowledge base [40]. Extensive discussions of load forecasting models can be found elsewhere [41].

In the middle of the 1990s, support vector regression (SVR) [42] began to be used to solve forecasting problems [43], and in the 2000s, Hong et al. [44–56] developed various SVR-based load forecasting models by hybridizing evolutionary algorithms, chaotic mapping functions and cloud theory with an SVR model, to effectively determine its three parameters to improve the forecasting accuracy. Based on Hong's research results, the accurate determination of three parameters of the SVR model is critical to improving its forecasting performance. The drawbacks of evolutionary algorithms cause the combination of parameters during the optimal modeling process, such as premature convergence or trapping in a local optimum. Therefore, Hong and his colleagues investigated the possibility of using chaotic mapping functions to increase the ergodicity over the search space, then transfer the three parameters into chaotic space to make the search more compact, and employ the cloud theory to establish a cooling mechanism during annealing process to enrich the influent effects of temperature decreasing mechanism, and eventually, improve the searching quality of SA algorithm for better forecasting accuracy.

Inspired by the excellent work of Hong et al., the authors find that the tabu search (TS) [57,58] algorithm is simply implemented to iteratively find a near-optimal solution, so it is powerful and has been successfully used to solve various optimization problems [59–61]. The TS algorithm, even with a flexible memory system to record recently visited solutions, and the ability to climb out of local minima, suffers from the tuning of the tabu tenure, meaning that it still becomes stuck at local minima and has a low convergence speed [62,63]. Also, the best solution is fixed for long iterations, i.e., it takes a great deal of time to escape to near-global optima from current position [64]. Therefore, both intensification and diversification strategies should be considered to improve the robustness, effectiveness and efficiency of simple TS; a more powerful neighborhood structure can be feasibly constructed by applying quantum computing concepts [65]. The same old problem, premature convergence or trapping at local optima, causes the forecasting accuracy to be unsatisfactory. This paper seeks to extend Hong's exploration to overcome the shortcomings of the TS algorithm, and to use the improved TS algorithm to forecast electric loads.

In this work, quantum computing concepts are utilized to improve the intensification and diversification of the simple TS algorithm; to improve its searching performance, and thus to improve its forecasting accuracy. The forecasting performance of the proposed hybrid quantum TS algorithm with an SVR model—the support vector regression quantum tabu search (SVRQTS) model—is compared with that of four other forecasting methods that were proposed by Hong [56] and Huang [66]. This paper is organized as follows. Section 2 presents the detail processes of the proposed SVRQTS model. The basic formulation of SVR and the quantum tabu search (QTS) algorithm are introduced. Section 3 presents two numerical examples and compares published methods with respect to forecasting accuracy. Finally, Section 4 draws conclusions.

2. Methodology of Support Vector Regression Quantum Tabu Search (SVRQTS) Model

2.1. Support Vector Regression (SVR) Model

A brief introduction of an SVR model is provided as follows. For a given training data set, $G = \{(x_i, y_i)\}_{i=1}^{n}$, where x_i is a vector of fed-in data and y_i is the corresponding actual values. G is then mapped into a high dimensional feature space by a nonlinear mapping function, $\varphi(\cdot)$. Theoretically, in the feature space, there should be an optimized linear function, f, to approximate the relationship between x_i and y_i. This kind of optimized linear function is the so-called SVR function and is shown as Equation (1),

$$f(x) = w^T \varphi(x) + b \tag{1}$$

where $f(x)$ represents the forecasting values; the coefficients w and b are coefficients which are estimated by minimizing the empirical risk function as shown in Equation (2),

$$R(f) = \frac{1}{N} \sum_{i=1}^{N} L_\varepsilon(y_i, w^T \varphi(x) + b) + \frac{1}{2} w^T w$$

$$L_\varepsilon(y, f(x)) = \begin{cases} |f(x) - y| - \varepsilon, & \text{if} \quad |f(x) - y| \geq \varepsilon \\ 0, & \text{otherwise} \end{cases} \tag{2}$$

where $L_\varepsilon(y, f(x))$ is the ε-insensitive loss function. The ε-insensitive loss function is employed to find out an optimum hyper plane on the high dimensional feature space to maximize the distance separating the training data into two subsets. Thus, the SVR focuses on finding the optimum hyper plane and minimizing the training error between the training data and the ε-insensitive loss function. The SVR model then minimizes the overall errors as shown in Equation (3),

$$\underset{w,b,\xi^*,\xi}{\text{Min}} \quad R(w, \xi^*, \xi) = \frac{1}{2} w^T w + C \sum_{i=1}^{N} (\xi_i^* + \xi_i) \text{ with the constraints :}$$

$$\begin{aligned} y_i - w^T \varphi(x_i) - b &\leq \varepsilon + \xi_i^*, & i = 1, 2, ..., N \\ -y_i + w^T \varphi(x_i) + b &\leq \varepsilon + \xi_i, & i = 1, 2, ..., N \\ \xi_i^* &\geq 0, & i = 1, 2, ..., N \\ \xi_i &\geq 0, & i = 1, 2, ..., N \end{aligned} \tag{3}$$

The first term of Equation (3), by employing the concept of maximizing the distance of two separated training data, is used to regularize weight sizes, to penalize large weights, and to maintain regression function flatness. The second term, to penalize the training errors of $f(x)$ and y, decides the balance between confidence risk and experience risk by using the ε-insensitive loss function. C is a parameter to specify the trade-off between the empirical risk and the model flatness. Training errors above ε are denoted as ξ_i^*, whereas training errors below $-\varepsilon$ are denoted as ξ_i, which are two positive slack variables, representing the distance from actual values to the corresponding boundary values of ε-tube.

After the quadratic optimization problem with inequality constraints is processed, the parameter vector **w** in Equation (1) is obtained in Equation (4),

$$w = \sum_{i=1}^{N} (\beta_i^* - \beta_i) \varphi(x_i) \tag{4}$$

where β_i^*, β_i, satisfying the equality $\beta_i * \beta_i^* = 0$, are the Lagrangian multipliers. Finally, the SVR regression function is obtained as Equation (5) in the dual space,

$$f(x) = \sum_{i=1}^{N} (\beta_i^* - \beta_i) K(x_i, x) + b \tag{5}$$

where $K(x_i, x_j)$ is so-called the kernel function, and the value of the kernel equals the inner product of two vectors, x_i and x_j, in the feature space $\varphi(x_i)$ and $\varphi(x_j)$, respectively; that is, $K(x_i, x_j) = \varphi(x_i) \cdot \varphi(x_j)$. However, the computation of the inner product in the high feature space becomes a computationally complicated problem along with the increase in the input dimensions. Such a problem of contradiction between high dimensions and computational complexity can be overcome by using the kernel trick or defining appropriate kernel functions in place of the dot product of the input vectors in high-dimensional feature space. The kernel function is used to directly compute the inner product from the input space, rather than in the high dimensional feature space. Kernel functions provide a way to avoid the curse of dimensionality. There are several types of kernel function, and it is hard to determine the type of kernel functions for specific data patterns [67]. The most commonly used kernel functions include linear functions, polynomial functions, Gaussian functions, sigmoid functions, splines, etc. The Gaussian function, $K(x_i, x_j) = \exp\left(-0.5\|x_i - x_j\|^2/\sigma^2\right)$, is used widely among all these various kernel functions as it can map the input space sample set into a high dimensional feature space effectively and is good for representing the complex nonlinear relationship between the input and output samples. Furthermore, only one variable (the width parameter, σ) is there to be defined. Considering the above advantages, the Gaussian radial basis function (RBF) is employed as the kernel function in this study.

The most important consideration in maximizing the forecasting accuracy of an SVR model is the well determination of its three parameters, which are the hyper-parameters, C, ε, and the kernel parameter, σ. Therefore, finding efficient algorithms for evaluating these three parameters is critical. As indicated above, inspired by Hong's hybridization of chaotic mapping functions with evolutionary algorithms to find favorable combinations of parameters and to overcome the premature convergence of the evolutionary algorithms, this work uses another (quantum-based) method to find an effective hybrid algorithm without the drawbacks of the TS algorithm by, for example, improving its intensification and diversification. Accordingly, the QTS algorithm is developed and improved using the hybrid chaotic mapping function. The chaotic QTS (CQTS) algorithm is hybridized with an SVR model, to develop the support vector regression chaotic quantum tabu search (SVRCQTS) model, to optimize parameter selection to maximize forecasting accuracy.

2.2. Chaotic Quantum Tabu Search Algorithm

2.2.1. Tabu Search (TS) Algorithm and Quantum Tabu Search (QTS) Algorithm

In 1986, Glover and Laguna first developed a renowned meta-heuristic algorithm called tabu Search (TS) [57,58]. TS is an iterative procedure designed for exploring in the solution space to find the near optimal solution. TS starts with a random solution or a solution obtained by a constructive and deterministic method and evaluates the fitness function. Then all possible neighbors of the given solution are generated and evaluated. A neighbor is a solution which can be reached from the current solution by a simple move. New solution is generated from the neighbors of the current one. To avoid retracing the used steps, the method records recent moves in a tabu list. The tabu list keeps track of previously explored solutions and forbids the search from returning to a previously visited solution. If the best of these neighbors is not in the tabu list, pick it to be the new current solution. One of the most important features of TS is that a new solution may be accepted even if the best neighbor solution is worse than the current one. In this way it is possible to overcome trapping in local minima. TS algorithm has been successfully used to lots of optimization problems [59–61].

However, in the TS algorithm, if a neighboring solution is not in the tabu list, TS sets it as the new current solution, but this solution is commonly worse than the current best solution. TS typically finds local minima and so do not change the best solution for many iterations; therefore, reaching a near-global minimum takes a long time and its convergence speed is low [62]. To overcome this shortcoming of the TS algorithm; to reduce its convergence time, to solve the similar old problem, premature convergence or trapping at local optima, the qubit concept and the quantum rotation gate mechanism can be used to construct a more powerful neighborhood structure by quantum computing concepts [65].

In the traditional TS algorithm, an initial solution is randomly generated, and its fitness function is evaluated to determine whether it should be set as the current best solution. However, in quantum computing, the initial solution is generated by using the concept of qubit to assign a real value in the interval (0,1), consistent with Equation (6). A qubit is the smallest unit of information for a quantum representation, and is mathematically represented as a column vector (unit vector), which can be identified in 2D Hilbert space. Equation (6) describes a quantum superposition between these two states. In quantum measurement, the super-position between states collapses into either the "ground state" or the "excited state".

$$|\psi\rangle = c_1 |0\rangle + c_2 |1\rangle \tag{6}$$

where $|0\rangle$ represents the "ground state", $|1\rangle$ denotes the "excited state"; $(c_1, c_2) \in \aleph$; c_1 and c_2 are the probability amplitudes of these two states; $\aleph$ is the set of complex numbers.

The most popular quantum gate, the quantum rotation gate (given by Equation (7)), is used to update the initial solution.

$$\begin{bmatrix} \alpha'_i \\ \beta'_i \end{bmatrix} = \begin{bmatrix} \cos(\theta_i) - \sin(\theta_i) \\ \sin(\theta_i) - \cos(\theta_i) \end{bmatrix} \begin{bmatrix} \alpha_i \\ \beta_i \end{bmatrix} \tag{7}$$

where (α'_i, β'_i) is the updated qubit; θ_i is the rotation angle.

The quantum orthogonality process (Equation (8)) is implemented to ensure that the corresponding value exceeds $rand(0,1)$. The tabu memory is introduced and set to null before the process is executed. The QTS begins with a single vector, v_{best}, and terminates when it reaches the predefined number of iterations. In each iteration, a new set of vectors, $V(BS)$ is generated in the neighborhood of v_{best}. For each vector in $V(BS)$, if it is not in the tabu memory and has a higher fitness value than v_{best}, then v_{best} is updated as the new vector. When the tabu memory is full, the first-in-first-out (FIFO) rule is applied to eliminate a vector from the list.

$$|c_1|^2 + |c_2|^2 = 1 \tag{8}$$

where $|c_1|^2$ and $|c_2|^2$ are the two probabilities that are required to transform the superposition between the states (as in Equation (6)) into $|0\rangle$ and $|1\rangle$, respectively.

2.2.2. Chaotic Mapping Function for Quantum Tabu Search (QTS) Algorithm

As mentioned, the chaotic variable can be adopted by applying the chaotic phenomenon to maintain diversity in the population to prevent premature convergence. The CQTS algorithm is based on the QTS algorithm, but uses the chaotic strategy when premature convergence occurs during the iterative searching process; at other times, the QTS algorithm is implemented, as described in Section 2.2.1.

To strengthening the effect of chaotic characteristics, many studies have used the logistic mapping function as a chaotic sequence generator. The greatest disadvantage of the logistic mapping function is that its distribution is concentration at both ends, with little in the middle. The Cat mapping function has a better chaotic distribution characteristic, so in this paper, the Cat mapping function is used as the chaotic sequence generator.

The classical Cat mapping function is the two-dimensional Cat mapping function [68], shown as Equation (9),

$$\begin{cases} x_{n+1} = (x_n + y_n) \bmod 1 \\ y_{n+1} = (x_n + 2y_n) \bmod 1 \end{cases} \tag{9}$$

where $x \bmod 1 = x - [x]$, mod, the so-called modulo operation, is used for the fractional parts of a real number x by subtracting an appropriate integer.

2.2.3. Implementation Steps of Chaotic Quantum Tabu Search (CQTS) Algorithm

The procedure of the hybrid CQTS algorithm with an SVR model is as follows; Figure 1 presents the corresponding flowchart.

Step 1 Initialization. Randomly generate the initial solution, P, that includes the values of three parameters in an SVR model.

Step 2 Objective value. Compute the objective values (forecasting errors) by using the initial solution, P. The mean absolute percentage error ($MAPE$), given by Equation (10), is used to measure the forecasting errors.

$$MAPE = \frac{1}{N}\sum_{i=1}^{N}\left|\frac{y_i - f_i}{y_i}\right| \times 100\% \tag{10}$$

where N is the number of forecasting periods; y_i is the actual value in period i; f_i denotes the forecast value in period i.

Step 3 Generate neighbors. Using the qubit concept, Equation (6) sets the initial solution, P, to a real value between interval (0,1) and then obtains P'. Then, use the quantum rotation gate, given by Equation (7), to generate the neighbor, P''.

Step 4 Pick. Pick a new individual from the examined neighbors based on the quantum tabu condition, which is determined by whether the corresponding value of P'' exceeds $rand(0,1)$.

Step 5 Update the best solution (objective value) and the tabu memory list. If $P'' > rand(0,1)$, then update the solution to P^* in the quantum tabu memory, v_{best}. Eventually, the objective value is updated as the current best solution. If the tabu memory is full, then the FIFO rule is applied to eliminate a P^* from the list.

Step 6 Premature convergence test. Calculate the mean square error (*MSE*), given by Equation (11), to evaluate the premature convergence status [69], and set the criteria, δ.

$$MSE = \frac{1}{S}\sum_{i=1}^{S}\left(\frac{f_i - f_{avg}}{f}\right)^2 \tag{11}$$

where f_i is the current objective value; f_{avg} is the mean of all previous objective values, and f is given by Equation (12),

$$f = \max\left\{1, \max_{\forall i \in S}\left\{|f_i - f_{avg}|\right\}\right\} \tag{12}$$

An *MSE* of less than δ indicates premature convergence. Therefore, the Cat mapping function, Equation (9), is used to find new optima, and the new optimal value is set as the best solution.

Step 7 Stopping criteria. If the stopping threshold (*MAPE*, which quantifies the forecasting accuracy) or the maximum number of iterations is reached, then training is stopped and the results output; otherwise, the process returns to **step 3**.

3. Numerical Examples

3.1. Data Set of Numerical Examples

3.1.1. The First Example: Taiwan Regional Load Data

In the first example, Taiwan's regional electricity load data from a published paper [56,66] are used to establish the proposed SVRCQTS forecasting model. The forecasting performances of this proposed model is compared with that of alternatives. The data set comprises 20 years (from 1981 to 2000) of load values for four regions of Taiwan. This data set is divided into several subsets—a training set (comprising 12 years of load data from 1981 to 1992), a validation set (comprising four years of data from 1993 to 1996), and a testing set (comprising four years of data from 1997 to 2000). The forecasting performances are measured using *MAPE* (Equation (10)).

In the training stage, the rolling forecasting procedure, proposed by Hong [56], is utilized to help CQTS algorithm determining appropriate parameter values of an SVR model in the training stage, and eventually, receive more satisfied results. For details, the training set is further divided into two subsets, namely the fed-in (for example, n load data) and the fed-out ($12 - n$ load data), respectively. Firstly, the preceding n load data are used to minimize the training error by the structural risk principle; then, receive one-step-ahead (in-sample) forecasting load, i.e., the $(n + 1)$th forecasting load. Secondly, the next n load data, i.e., from 2nd to $(n + 1)$th data, are set as the new fed-in and similarly used to minimize the training error again to receive the second one-step-ahead (in-sample) forecasting load, named as the $(n + 2)$th forecasting load. Repeat this procedure until the 12nd (in-sample) forecasting load is obtained with the training error. The training error can be obtained during each iteration, these parameters would be decided by QTS algorithm, and the validation error would be also calculated in the meanwhile. Only with the smallest validation and testing errors will the adjusted parameter combination be selected as the most appropriate parameter combination. The testing data set is only used for examining the forecasting accuracy level. Eventually, the four-year forecasting electricity load demands in each region are forecasted by the SVRCQTS model. The complete process is illustrated in Figure 2.

3.1.2. The Second Example: Taiwan Annual Load Data

In the second example, Taiwan's annual electricity load data from a published paper are used [56,66]. The data set is composed of 59 years of load data (from 1945 to 2003), which are divided into three subsets—a training set (40 years of load data from 1945 to 1984), a validation set (10 ten of load data from 1985 to 1994), and a testing set (nine years of load data from 1995 to 2003). The relevant modeling procedures are as in the first example.

3.1.3. The Third Example: 2014 Global Energy Forecasting Competition (GEFCOM 2014) Load Data

The third example involves the 744 h of load data from the 2014 Global Energy Forecasting Competition [70] (from 00:00 1 December 2011 to 00:00 1 January 2012). The data set is divided into three subsets—a training set (552 h of load data from 01:00 1 December 2011 to 00:00 24 December 2011), a validation set (96 h of load data from 01:00 24 December 2011 to 00:00 28 December 2011), and testing set (96 h of load data from 01:00 28 December 2011 to 00:00 1 January 2012). The relevant modeling procedures are as in the preceding two examples.

3.2. The SVRCQTS Load Forecasting Model

3.2.1. Parameters Setting in CQTS Algorithm

For some controlling parameters settings during modeling process, such as the total number of iteration is all fixed as 10,000; $\sigma \in [0, 15]$, $\varepsilon \in [0, 100]$ in all examples, $C \in [0, 20,000]$ in Example 1, $C \in [0, 3 \times 10^{10}]$ in Examples 2 and 3; δ is all set to 0.001.

3.2.2. Forecasting Results and Analysis for Example 1

In Example 1, the combination of parameters of the most appropriate model are evaluated using the QTS algorithm and the CQTS algorithm for each region, and almost has the smallest testing *MAPE* value. Table 1 presents these well-determined parameters for each region.

Table 1. Parameters determination of SVRCQTS and SVRQTS models (example 1). SVRCQTS: support vector regression chaotic quantum tabu search; SVRQTS: support vector regression quantum Tabu search.

Regions	SVRCQTS Parameters			*MAPE* of Testing (%)
	σ	C	ε	
Northern	10.0000	0.8000×10^{10}	0.7200	1.0870
Central	6.0000	1.6000×10^{10}	0.5500	1.2650
Southern	8.0000	1.4000×10^{10}	0.6500	1.1720
Eastern	8.0000	0.8000×10^{10}	0.4300	1.5430

Regions	SVRQTS Parameters			*MAPE* of Testing (%)
	σ	C	ε	
Northern	4.0000	0.8000×10^{10}	0.2500	1.3260
Central	12.0000	1.0000×10^{10}	0.2800	1.6870
Southern	10.0000	0.8000×10^{10}	0.7200	1.3670
Eastern	8.0000	1.4000×10^{10}	0.4200	1.9720

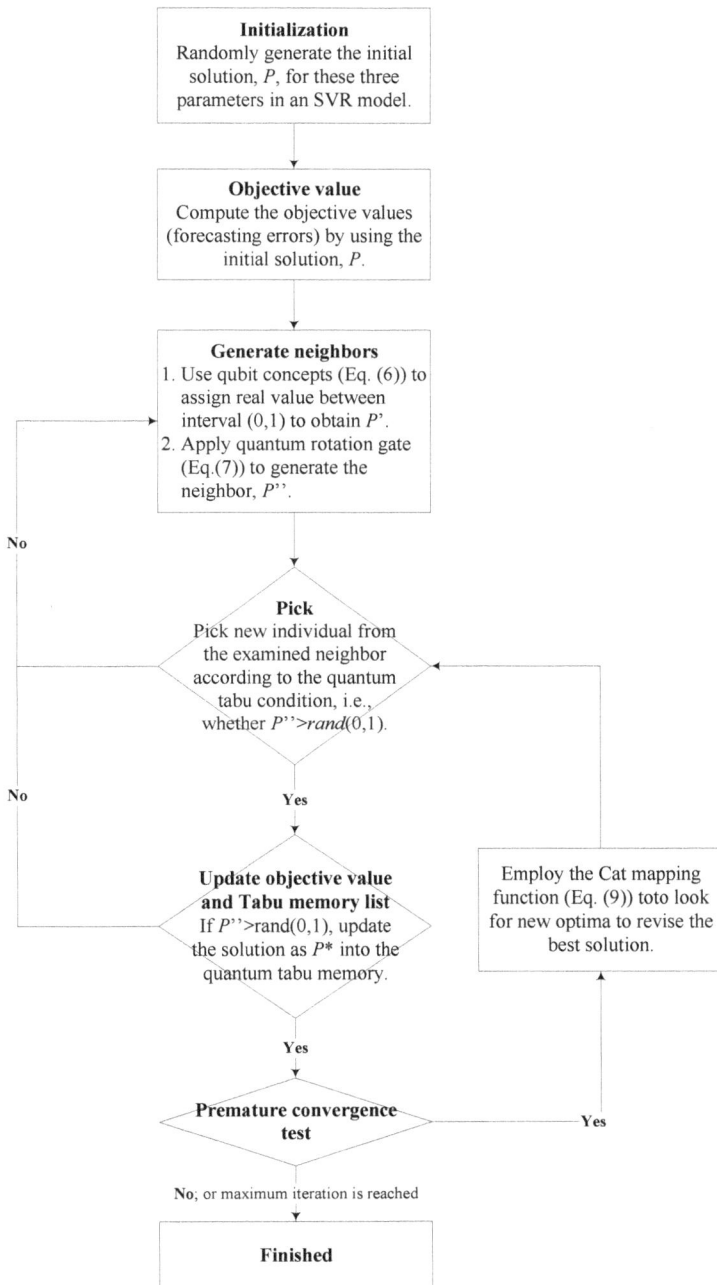

Figure 1. Quantum tabu search (QTS) algorithm flowchart.

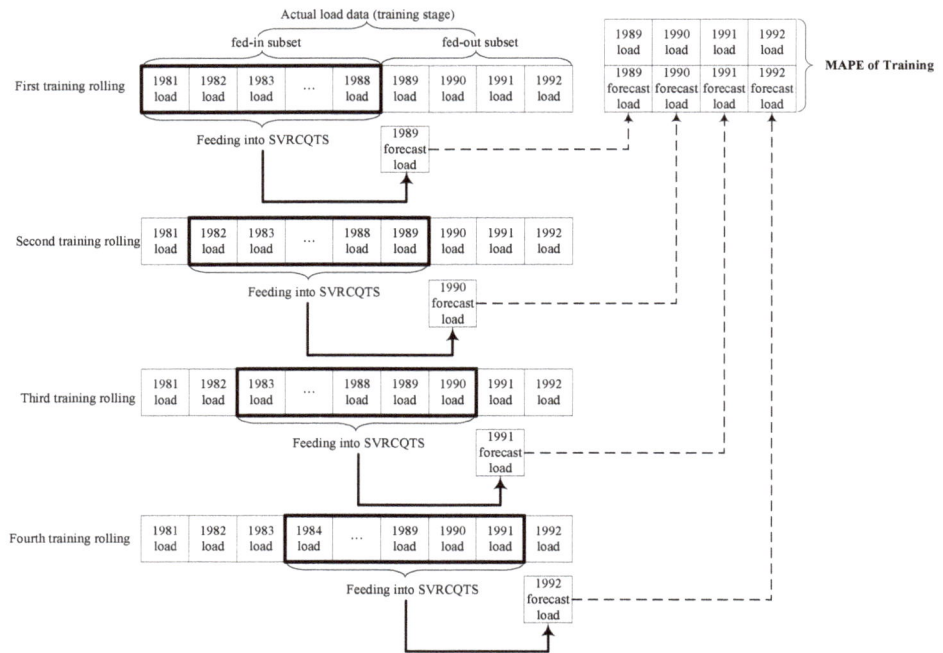

Figure 2. The rolling-based forecasting procedure.

Table 2 presents the forecasting accuracy index (*MAPE*) and electricity load values of each region that are forecast, under the same conditions, using alternative models, which were include SVRCQTS, SVRQTS, SVR with chaotic particle swarm optimization (SVRCQPSO), SVR with quantum PSO (SVRQPSO), and SVR with PSO (SVRPSO) models. Clearly, according to Table 2, the SVRCQTS model is superior to the other SVR-based models. Applying quantum computing mechanics to the TS algorithm is a feasible means of improving the satisfied, and thus improving the forecasting accuracy of the SVR model. The Cat mapping function has a critical role in finding an improved solution when the QTS algorithm becomes trapped in local optima or requires a long time to solve the problem of interest). For example, for the central region, the QTS algorithm is utilized to find the best solution, $(\sigma, C, \varepsilon) = (12.0000, 1.0000 \times 1010, 0.2800)$, with a forecasting error, *MAPE*, of 1.6870%. The solution can be further improved by using the CQTS algorithm with $(\sigma, C, \varepsilon) = (6.0000, 1.6000 \times 1010, 0.5500)$, which has a smaller forecasting accuracy of 1.2650%. For other regions, the QTS algorithm yields an increased forecasting performance to 1.3260% (northern region), 1.3670% (southern region) and 1.9720% (eastern region). All of these models can also be further improved to increase the accuracy of the forecasting results by using the Cat mapping function (the CQTS algorithm), yielding a forecasting accuracy of 1.087% for the northern region, 1.1720% for the southern region, and 1.5430% for the eastern region.

To verify that the proposed SVRCQTS and SVRQTS models offers an improved forecasting accuracy, the Wilcoxon signed-rank test, recommended by Diebold and Mariano [71], is used. In this work, the Wilcoxon signed-rank test is performed with two significance levels, $\alpha = 0.025$ and $\alpha = 0.005$, by one-tail-tests. Table 3 presents the test results, which reveal that the SVRCQTS model significantly outperforms other models for the northern and eastern regions in terms of *MAPE*.

Table 2. Forecasting results (and absolute errors) of SVRCQTS, SVRQTS, and other models (example 1) (unit: 10^6 Wh).

Year	Northern Region					
	SVRCQTS	SVRQTS	SVRCQPSO	SVRQPSO	SVRCPSO	SVRPSO
1997	11,123 (101)	11,101 (121)	11,339 (117)	11,046 (176)	11,232 (10)	11,245 (23)
1998	11,491 (151)	11,458 (184)	11,779 (137)	11,787 (145)	11,628 (14)	11,621 (21)
1999	12,123 (142)	12,154 (173)	11,832 (149)	12,144 (163)	12,016 (35)	12,023 (42)
2000	13,052 (128)	13,080 (156)	12,798 (126)	12,772 (152)	12,306 (618)	12,306 (618)
MAPE (%)	**1.0870**	1.3260	1.1070	1.3370	1.3187	1.3786

Year	Central Region					
	SVRCQTS	SVRQTS	SVRCQPSO	SVRQPSO	SVRCPSO	SVRPSO
1997	5009 (52)	5132 (71)	4987 (74)	5140 (79)	5066 (5)	5085 (24)
1998	5167 (79)	5142 (104)	5317 (71)	5342 (96)	5168 (78)	5141 (105)
1999	5301 (68)	5318 (85)	5172 (61)	5130 (103)	5232 (1)	5236 (3)
2000	5702 (69)	5732 (99)	5569 (64)	5554 (79)	5313 (320)	5343 (290)
MAPE (%)	**1.2650**	1.6870	1.2840	1.6890	1.8100	1.9173

Year	Southern Region					
	SVRCQTS	SVRQTS	SVRCQPSO	SVRQPSO	SVRCPSO	SVRPSO
1997	6268 (68)	6436 (100)	6262 (74)	6265 (71)	6297 (39)	6272 (64)
1998	6398 (80)	6245 (73)	6401 (83)	6418 (100)	6311 (7)	6314 (4)
1999	6343 (84)	6338 (79)	6179 (80)	6178 (81)	6324 (65)	6327 (68)
2000	6735 (69)	6704 (100)	6738 (66)	6901 (97)	6516 (288)	6519 (285)
MAPE (%)	**1.1720**	1.3670	1.1840	1.3590	1.4937	1.5899

Year	Eastern Region					
	SVRCQTS	SVRQTS	SVRCQPSO	SVRQPSO	SVRCPSO	SVRPSO
1997	362 (4)	364 (6)	353 (5)	350 (8)	370 (12)	367 (9)
1998	390 (7)	388 (9)	404 (7)	390 (7)	376 (21)	374 (23)
1999	395 (6)	394 (7)	394 (7)	410 (9)	411 (10)	409 (8)
2000	427 (7)	429 (9)	414 (6)	413 (7)	418 (2)	415 (5)
MAPE (%)	**1.5430**	1.9720	1.5940	1.9830	2.1860	2.3094

Note: *: The values in the parentheses are the absolute error, which is defined as: $|y_i - f_i|$, where y_i is the actual value in period i; f_i denotes is the forecast value in period i. SVRCQPSO: support vector regression chaotic quantum particle swarm optimization; SVRQPSO: support vector regression quantum particle swarm optimization ; SVRCPSO: support vector regression chaotic particle swarm optimization; SVRPSO: support vector regression particle swarm optimization.

Table 3. Wilcoxon signed-rank test (example 1).

Compared Models		Wilcoxon Signed-Rank Test		
		$\alpha = 0.025; W = 0$	$\alpha = 0.005; W = 0$	*p*-Value
Northern region	SVRCQTS vs. SVRPSO	0 [a]	0 [a]	N/A
	SVRCQTS vs. SVRCPSO	0 [a]	0 [a]	N/A
	SVRCQTS vs. SVRQPSO	0 [a]	0 [a]	N/A
	SVRCQTS vs. SVRCQPSO	0 [a]	0 [a]	N/A
	SVRCQTS vs. SVRQTS	1	1	N/A
Central region	SVRCQTS vs. SVRPSO	0 [a]	0 [a]	N/A
	SVRCQTS vs. SVRCPSO	0 [a]	0 [a]	N/A
	SVRCQTS vs. SVRQPSO	0 [a]	0 [a]	N/A
	SVRCQTS vs. SVRCQPSO	1	1	N/A
	SVRCQTS vs. SVRQTS	0 [a]	0 [a]	N/A
Southern region	SVRCQTS vs. SVRPSO	0 [a]	0 [a]	N/A
	SVRCQTS vs. SVRCPSO	0 [a]	0 [a]	N/A
	SVRCQTS vs. SVRQPSO	0 [a]	0 [a]	N/A
	SVRCQTS vs. SVRCQPSO	1	1	N/A
	SVRCQTS vs. SVRQTS	0 [a]	0 [a]	N/A
Eastern region	SVRCQTS vs. SVRPSO	0 [a]	0 [a]	N/A
	SVRCQTS vs. SVRCPSO	0 [a]	0 [a]	N/A
	SVRCQTS vs. SVRQPSO	0 [a]	0 [a]	N/A
	SVRCQTS vs. SVRCQPSO	1	1	N/A
	SVRCQTS vs. SVRQTS	0 [a]	0 [a]	N/A

Note: [a] denotes that the SVRCQTS model significantly outperforms other alternative models.

3.2.3. Forecasting Results and Analysis for Example 2

In Example 2, the processing steps are those in the preceding example. The parameters in an SVR model are computed using the QTS algorithm and the CQTS algorithm. The finalized models exhibit the best forecasting performance with the smallest *MAPE* values. Table 4 presents the well determined parameters for annual electricity load data. To compare with other benchmarking algorithms, Table 4 presents all results from relevant papers on SVR-based modeling, such as those of Hong [56], who proposed the SVRCPSO and SVR with PSO (SVRPSO) models and Huang [66], who proposed the SVRCQPSO and SVRQPSO models.

Table 5 presents the *MAPE* values and forecasting results obtained using the alternative forecasting models. The SVRCQTS model outperforms the other models, indicating quantum computing is an ideal approach to improve the performance of any SVR-based model, and that the Cat mapping function is very effective for solving the problem of premature convergence and the fact that it is time-saving. Clearly, the QTS algorithm yields $(\sigma, C, \varepsilon) = (5.0000, 1.3000 \times 10^{11}, 0.630)$ with a *MAPE* of 1.3210%, whereas the CQTS algorithm provides a better solution, $(\sigma, C, \varepsilon) = (6.0000, 1.8000 \times 10^{11}, 0.340)$ with a *MAPE* of 1.1540%. Figure 3 presents the real values and the forecast values obtained using the various models.

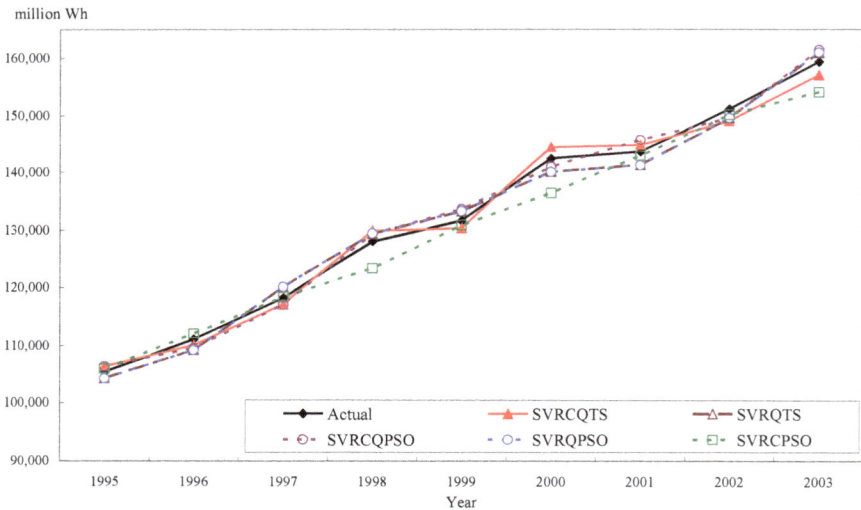

Figure 3. Actual values and forecasting values of SVRCQTS, SVRQTS, and other models (Example 2). SVRCQTS: support vector regression chaotic quantum Tabu search; SVRQTS: support vector regression quantum Tabu search.

Table 4. Parameters determination of SVRCQTS and SVRQTS models (Example 2). PSO: particle swarm optimization; CPSO: chaotic particle swarm optimization; QPSO: quantum particle swarm optimization; CQPSO: chaotic quantum particle swarm optimization.

Optimization Algorithms	Parameters			MAPE of Testing (%)
	σ	C	ε	
PSO algorithm [56]	0.2293	1.7557×10^{11}	10.175	3.1429
CPSO algorithm [56]	0.2380	2.3365×10^{11}	39.296	1.6134
QPSO algorithm [66]	12.0000	0.8000×10^{11}	0.380	1.3460
CQPSO algorithm [66]	10.0000	1.5000×10^{11}	0.560	1.1850
QTS algorithm	5.0000	1.3000×10^{11}	0.630	1.3210
CQTS algorithm	6.0000	1.8000×10^{11}	0.340	1.1540

Table 5. Forecasting results (and absolute errors) of SVRCQTS, SVRQTS, and other models (unit: 10^6 Wh).

Years	SVRCQTS	SVRQTS	SVRCQPSO	SVRQPSO	SVRCPSO	SVRPSO
1995	106,353 (985)	104,241 (1127)	106,379 (1011)	104,219 (1149)	105,960 (592)	102,770 (2598)
1996	110,127 (1013)	109,246 (1894)	109,573 (1567)	109,210 (1930)	112,120 (980)	109,800 (1340)
1997	117,180 (1119)	120,174 (1875)	117,149 (1150)	120,210 (1911)	118,450 (151)	115,570 (2729)
1998	130,023 (1893)	129,501 (1371)	129,466 (1336)	129,527 (1397)	123,400 (4730)	120,650 (7480)
1999	130,464 (1262)	133,275 (1549)	133,646 (1920)	133,304 (1578)	130,940 (786)	128,240 (3486)
2000	144,500 (2087)	140,099 (2314)	140,945 (1468)	140,055 (2358)	136,420 (5993)	137,250 (5163)
2001	144,884 (1260)	141,271 (2353)	145,734 (2110)	141,227 (2397)	142,910 (714)	140,230 (3394)
2002	149,099 (2094)	149,675 (1518)	149,652 (1541)	149,646 (1547)	150,210 (983)	151,150 (43)
2003	157,099 (2281)	161,001 (1621)	161,458 (2078)	161,032 (1652)	154,130 (5250)	146,940 (12,440)
MAPE (%)	**1.1540**	**1.3210**	**1.1850**	**1.3460**	**1.6134**	**3.1429**

Note: *: the value in the parentheses is the absolute error which is defined as: $|y_i - f_i|$, where y_i is the actual value in period i; f_i denotes is the forecast value in period i.

To ensure the significance of the proposed SVRCQTS model in Example 2, the Wilcoxon signed-rank test is again considered performed. Table 6 shows that the SVRCQTS model passes the test and significantly improves on other alternatives.

Table 6. Wilcoxon signed-rank test (example 2).

Compared Models	Wilcoxon Signed-Rank Test		
	$\alpha = 0.025; W = 5$	$\alpha = 0.005; W = 8$	*p*-Value
SVRCQTS vs. SVRPSO	2 [a]	2 [a]	N/A
SVRCQTS vs. SVRCPSO	3 [a]	3 [a]	N/A
SVRCQTS vs. SVRQPSO	4 [a]	4 [a]	N/A
SVRCQTS vs. SVRCQPSO	4 [a]	4 [a]	N/A
SVRCQTS vs. SVRQTS	4 [a]	4 [a]	N/A

Note: [a] denotes that the SVRCQTS model significantly outperforms other alternative models.

3.2.4. Forecasting Results and Analysis for Example 3

In Example 3, the modeling processes are the same as in the preceding two examples. The parameters in the SVR model are calculated using the QTS algorithm and the CQTS algorithm. Table 7 presents the details of the determined models and the alternatives models for the GEFCOM 2014 data set. Huang [66] used the GEFCOM 2014 load data to forecast the electric load, therefore, the models of Huang [66] are included in this paper as the alternative models.

The alternative models of Huang [66]—ARIMA(0,1,1), back propagation neural networks (BPNN), SVRPSO, SVRCPSO, SVRQPSO, SVRCQPSO, and SVRQTS models—are compared herein under fixed conditions. Figure 4 displays the real values and the forecast results obtained using all compared models, and demonstrates that the SVRCQTS model outperforms the SVRQTS and SVRQPSO models. It also reveals that applying quantum computing mechanics to the TS algorithm improves the forecasting accuracy level for any SVR-based models. The SVRCQTS model has a smaller MAPE value than the SVRQTS model.

Table 7. Parameters determination of SVRCQTS, SVRQTS, and other models (Example 3).

Optimization Algorithms	Parameters			*MAPE* of Testing (%)
	σ	C	ε	
PSO algorithm [66]	7.000	34.000	0.9400	3.1500
CPSO algorithm [66]	22.000	19.000	0.6900	2.8600
QPSO algorithm [66]	9.000	42.000	0.1800	1.9600
CQPSO algorithm [66]	19.000	35.000	0.8200	1.2900
QTS algorithm	25.000	67.000	0.0900	1.8900
CQTS algorithm	12.000	26.000	0.3200	1.3200

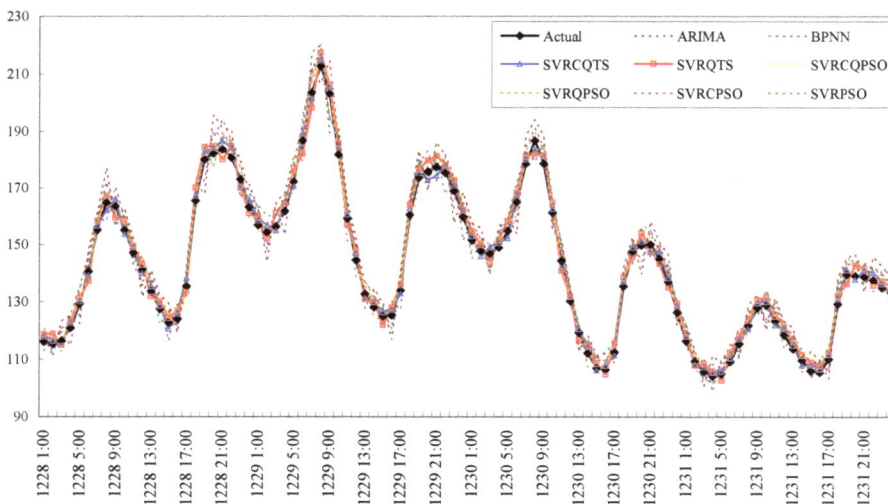

Figure 4. Actual values and forecasting values of SVRCQTS, SVRQTS, and other models (Example 3).

Finally, Table 8 presents the results of the Wilcoxon signed-rank test. It indicates that the proposed SVRCQTS model almost receives statistical significance in forecasting performances under the significant level, $\alpha = 0.05$. Therefore, the proposed SVRCQTS model significantly outperforms other alternatives in terms of $\alpha = 0.05$.

Table 8. Wilcoxon signed-rank test (Example 3).

Compared Models	Wilcoxon Signed-Rank Test		
	$\alpha = 0.025; W = 2328$	$\alpha = 0.05; W = 2328$	p-Value
SVRCQTS vs. ARIMA	1621 [a]	1621 [a]	0.00988
SVRCQTS vs. BPNN	1600 [a]	1600 [a]	0.00782
SVRCQTS vs. SVRPSO	2148	2148 [a]	0.04318
SVRCQTS vs. SVRCPSO	2163	2163 [a]	0.04763
SVRCQTS vs. SVRQPSO	1568 [a]	1568 [a]	0.00544
SVRCQTS vs. SVRCQPSO	1344 [a]	1344 [a]	0.00032
SVRCQTS vs. SVRQTS	1741	1741 [a]	0.03156

Note: [a] denotes that the SVRCQTS model significantly outperforms other alternative models.

4. Conclusions

This work proposes a hybrid model that incorporates an SVR-based model, the chaotic cat mapping function, and the QTS algorithm for forecasting electricity load demand. Experimental results reveal that the proposed model exhibits significantly better forecasting performance than other SVR-based forecasting models. In this paper, quantum mechanics is utilized to improve the intensification and diversification of the simple TS algorithm, and thereby to improve its forecasting accuracy. Chaotic cat mapping is also used to help prevent the QTS algorithm from becoming trapped in local optima in the modeling processes. This work marks a favorable beginning of the hybridization of quantum computing mechanics and the chaotic mechanism to expand the search space, which is typically limited by Newtonian dynamics.

Author Contributions: Bing-Yi Lin designed the experiments, analyzed the data, and wrote the paper. Cheng-Wen Lee provided professional guidance.

Conflicts of Interest: The author declares no conflict of interest.

References

1. Bunn, D.W.; Farmer, E.D. Comparative models for electrical load forecasting. *Int. J. Forecast.* **1986**, *2*, 241–242.
2. Fan, S.; Chen, L. Short-term load forecasting based on an adaptive hybrid method. *IEEE Trans. Power Syst.* **2006**, *21*, 392–401. [CrossRef]
3. Morimoto, R.; Hope, C. The impact of electricity supply on economic growth in Sri Lanka. *Energy Econ.* **2004**, *26*, 77–85. [CrossRef]
4. Zhao, W.; Wang, J.; Lu, H. Combining forecasts of electricity consumption in China with time-varying weights updated by a high-order Markov chain model. *Omega* **2014**, *45*, 80–91. [CrossRef]
5. Wang, J.; Wang, J.; Li, Y.; Zhu, S.; Zhao, J. Techniques of applying wavelet de-noising into a combined model for short-term load forecasting. *Int. J. Electr. Power Energy Syst.* **2014**, *62*, 816–824. [CrossRef]
6. Hussain, A.; Rahman, M.; Memon, J.A. Forecasting electricity consumption in Pakistan: The way forward. *Energy Policy* **2016**, *90*, 73–80. [CrossRef]
7. Maçaira, P.M.; Souza, R.C.; Oliveira, F.L.C. Modelling and forecasting the residential electricity consumption in Brazil with pegels exponential smoothing techniques. *Procedia Comput. Sci.* **2015**, *55*, 328–335. [CrossRef]
8. Al-Hamadi, H.M.; Soliman, S.A. Short-term electric load forecasting based on Kalman filtering algorithm with moving window weather and load model. *Electr. Power Syst. Res.* **2004**, *68*, 47–59. [CrossRef]
9. Zheng, T.; Girgis, A.A.; Makram, E.B. A hybrid wavelet-Kalman filter method for load forecasting. *Electr. Power Syst. Res.* **2000**, *54*, 11–17. [CrossRef]
10. Blood, E.A.; Krogh, B.H.; Ilic, M.D. Electric power system static state estimation through Kalman filtering and load forecasting. In Proceedings of the Power and Energy Society General Meeting—Conversion and Delivery of Electrical Energy in the 21st Century, Pittsburgh, PA, USA, 20–24 July 2008.
11. Hippert, H.S.; Taylor, J.W. An evaluation of Bayesian techniques for controlling model complexity and selecting inputs in a neural network for short-term load forecasting. *Neural Netw.* **2010**, *23*, 386–395. [CrossRef] [PubMed]
12. Zhang, W.; Yang, J. Forecasting natural gas consumption in China by Bayesian model averaging. *Energy Rep.* **2015**, *1*, 216–220. [CrossRef]
13. Chai, J.; Guo, J.E.; Lu, H. Forecasting energy demand of China using Bayesian combination model. *China Popul. Resour. Environ.* **2008**, *18*, 50–55.
14. Vu, D.H.; Muttaqi, K.M.; Agalgaonkar, A.P. A variance inflation factor and backward elimination based robust regression model for forecasting monthly electricity demand using climatic variables. *Appl. Energy* **2015**, *140*, 385–394. [CrossRef]
15. Dudek, G. Pattern-based local linear regression models for short-term load forecasting. *Electr. Power Syst. Res.* **2016**, *130*, 139–147. [CrossRef]
16. Wu, J.; Wang, J.; Lu, H.; Dong, Y.; Lu, X. Short term load forecasting technique based on the seasonal exponential adjustment method and the regression model. *Energy Convers. Manag.* **2013**, *70*, 1–9. [CrossRef]
17. Hernández, L.; Baladrón, C.; Aguiar, J.M.; Carro, B.; Sánchez-Esguevillas, A.; Lloret, J. Artificial neural networks for short-term load forecasting in microgrids environment. *Energy* **2014**, *75*, 252–264. [CrossRef]
18. Ertugrul, Ö.F. Forecasting electricity load by a novel recurrent extreme learning machines approach. *Int. J. Electr. Power Energy Syst.* **2016**, *78*, 429–435. [CrossRef]
19. Quan, H.; Srinivasan, D.; Khosravi, A. Uncertainty handling using neural network-based prediction intervals for electrical load forecasting. *Energy* **2014**, *73*, 916–925. [CrossRef]
20. Li, P.; Li, Y.; Xiong, Q.; Chai, Y.; Zhang, Y. Application of a hybrid quantized Elman neural network in short-term load forecasting. *Int. J. Electr. Power Energy Syst.* **2014**, *55*, 749–759. [CrossRef]
21. Sousa, J.C.; Neves, L.P.; Jorge, H.M. Assessing the relevance of load profiling information in electrical load forecasting based on neural network models. *Int. J. Electr. Power Energy Syst.* **2012**, *40*, 85–93. [CrossRef]
22. Lahouar, A.; Slama, J.B.H. Day-ahead load forecast using random forest and expert input selection. *Energy Convers. Manag.* **2015**, *103*, 1040–1051. [CrossRef]
23. Bennett, C.J.; Stewart, R.A.; Lu, J.W. Forecasting low voltage distribution network demand profiles using a pattern recognition based expert system. *Energy* **2014**, *67*, 200–212. [CrossRef]
24. Santana, Á.L.; Conde, G.B.; Rego, L.P.; Rocha, C.A.; Cardoso, D.L.; Costa, J.C.W.; Bezerra, U.H.; Francês, C.R.L. PREDICT—Decision support system for load forecasting and inference: A new undertaking for Brazilian power suppliers. *Int. J. Electr. Power Energy Syst.* **2012**, *38*, 33–45. [CrossRef]

25. Efendi, R.; Ismail, Z.; Deris, M.M. A new linguistic out-sample approach of fuzzy time series for daily forecasting of Malaysian electricity load demand. *Appl. Soft Comput.* **2015**, *28*, 422–430. [CrossRef]
26. Hooshmand, R.-A.; Amooshahi, H.; Parastegari, M. A hybrid intelligent algorithm based short-term load forecasting approach. *Int. J. Electr. Power Energy Syst.* **2013**, *45*, 313–324. [CrossRef]
27. Akdemir, B.; Çetinkaya, N. Long-term load forecasting based on adaptive neural fuzzy inference system using real energy data. *Energy Procedia* **2012**, *14*, 794–799. [CrossRef]
28. Chaturvedi, D.K.; Sinha, A.P.; Malik, O.P. Short term load forecast using fuzzy logic and wavelet transform integrated generalized neural network. *Int. J. Electr. Power Energy Syst.* **2015**, *67*, 230–237. [CrossRef]
29. Liao, G.-C. Hybrid Improved Differential Evolution and Wavelet Neural Network with load forecasting problem of air conditioning. *Int. J. Electr. Power Energy Syst.* **2014**, *61*, 673–682. [CrossRef]
30. Zhai, M.-Y. A new method for short-term load forecasting based on fractal interpretation and wavelet analysis. *Int. J. Electr. Power Energy Syst.* **2015**, *69*, 241–245. [CrossRef]
31. Zeng, X.; Shu, L.; Huang, G.; Jiang, J. Triangular fuzzy series forecasting based on grey model and neural network. *Appl. Math. Model.* **2016**, *40*, 1717–1727. [CrossRef]
32. Efendigil, T.; Önüt, S.; Kahraman, C. A decision support system for demand forecasting with artificial neural networks and neuro-fuzzy models: A comparative analysis. *Expert Syst. Appl.* **2009**, *36*, 6697–6707. [CrossRef]
33. Coelho, V.N.; Coelho, I.M.; Coelho, B.N.; Reis, A.J.R.; Enayatifar, R.; Souza, M.J.F.; Guimarães, F.G. A self-adaptive evolutionary fuzzy model for load forecasting problems on smart grid environment. *Appl. Energy* **2016**, *169*, 567–584. [CrossRef]
34. Khashei, M.; Mehdi Bijari, M. A novel hybridization of artificial neural networks and ARIMA models for time series forecasting. *Appl. Soft Comput.* **2011**, *11*, 2664–2675. [CrossRef]
35. Ghanbari, A.; Kazemi, S.M.R.; Mehmanpazir, F.; Nakhostin, M.M. A Cooperative Ant Colony Optimization-Genetic Algorithm approach for construction of energy demand forecasting knowledge-based expert systems. *Knowl.-Based Syst.* **2013**, *39*, 194–206. [CrossRef]
36. Bahrami, S.; Hooshmand, R.-A.; Parastegari, M. Short term electric load forecasting by wavelet transform and grey model improved by PSO (particle swarm optimization) algorithm. *Energy* **2014**, *72*, 434–442. [CrossRef]
37. Suykens, J.A.K.; Vandewalle, J.; De Moor, B. Optimal control by least squares support vector machines. *Neural Netw.* **2001**, *14*, 23–35. [CrossRef]
38. Aras, S.; Kocakoç, İ.D. A new model selection strategy in time series forecasting with artificial neural networks: IHTS. *Neurocomputing* **2016**, *174*, 974–987. [CrossRef]
39. Kendal, S.L.; Creen, M. *An Introduction to Knowledge Engineering*; Springer: London, UK, 2007.
40. Cherroun, L.; Hadroug, N.; Boumehraz, M. Hybrid approach based on ANFIS models for intelligent fault diagnosis in industrial actuator. *J. Control Electr. Eng.* **2013**, *3*, 17–22.
41. Hahn, H.; Meyer-Nieberg, S.; Pickl, S. Electric load forecasting methods: Tools for decision making. *Eur. J. Oper. Res.* **2009**, *199*, 902–907. [CrossRef]
42. Drucker, H.; Burges, C.C.J.; Kaufman, L.; Smola, A.; Vapnik, V. Support vector regression machines. *Adv. Neural Inf. Process Syst.* **1997**, *9*, 155–161.
43. Tay, F.E.H.; Cao, L.J. Modified support vector machines in financial time series forecasting. *Neurocomputing* **2002**, *48*, 847–861. [CrossRef]
44. Hong, W.C. Electric load forecasting by seasonal recurrent SVR (support vector regression) with chaotic artificial bee colony algorithm. *Energy* **2011**, *36*, 5568–5578. [CrossRef]
45. Chen, Y.H.; Hong, W.C.; Shen, W.; Huang, N.N. Electric load forecasting based on LSSVM with fuzzy time series and global harmony search algorithm. *Energies* **2016**, *9*, 70. [CrossRef]
46. Fan, G.; Peng, L.-L.; Hong, W.-C.; Sun, F. Electric load forecasting by the SVR model with differential empirical mode decomposition and auto regression. *Neurocomputing* **2016**, *173*, 958–970. [CrossRef]
47. Geng, J.; Huang, M.L.; Li, M.W.; Hong, W.C. Hybridization of seasonal chaotic cloud simulated annealing algorithm in a SVR-based load forecasting model. *Neurocomputing* **2015**, *151*, 1362–1373. [CrossRef]
48. Ju, F.Y.; Hong, W.C. Application of seasonal SVR with chaotic gravitational search algorithm in electricity forecasting. *Appl. Math. Model.* **2013**, *37*, 9643–9651. [CrossRef]
49. Fan, G.; Wang, H.; Qing, S.; Hong, W.C.; Li, H.J. Support vector regression model based on empirical mode decomposition and auto regression for electric load forecasting. *Energies* **2013**, *6*, 1887–1901. [CrossRef]

50. Hong, W.C.; Dong, Y.; Zhang, W.Y.; Chen, L.Y.; Panigrahi, B.K. Cyclic electric load forecasting by seasonal SVR with chaotic genetic algorithm. *Int. J. Electr. Power Energy Syst.* **2013**, *44*, 604–614. [CrossRef]
51. Zhang, W.Y.; Hong, W.C.; Dong, Y.; Tsai, G.; Sung, J.T.; Fan, G. Application of SVR with chaotic GASA algorithm in cyclic electric load forecasting. *Energy* **2012**, *45*, 850–858. [CrossRef]
52. Hong, W.C.; Dong, Y.; Lai, C.Y.; Chen, L.Y.; Wei, S.Y. SVR with hybrid chaotic immune algorithm for seasonal load demand forecasting. *Energies* **2011**, *4*, 960–977. [CrossRef]
53. Hong, W.C. Application of chaotic ant swarm optimization in electric load forecasting. *Energy Policy* **2010**, *38*, 5830–5839. [CrossRef]
54. Hong, W.C. Hybrid evolutionary algorithms in a SVR-based electric load forecasting model. *Int. J. Electr. Power Energy Syst.* **2009**, *31*, 409–417. [CrossRef]
55. Hong, W.C. Electric load forecasting by support vector model. *Appl. Math. Model.* **2009**, *33*, 2444–2454. [CrossRef]
56. Hong, W.C. Chaotic particle swarm optimization algorithm in a support vector regression electric load forecasting model. *Energy Convers. Manag.* **2009**, *50*, 105–117. [CrossRef]
57. Glover, F. Tabu search, part I. *ORSA J. Comput.* **1989**, *1*, 190–206. [CrossRef]
58. Glover, F. Tabu search, part II. *ORSA J. Comput.* **1990**, *2*, 4–32. [CrossRef]
59. Zeng, Z.; Yu, X.; He, K.; Huang, W.; Fu, Z. Iterated tabu search and variable neighborhood descent for packing unequal circles into a circular container. *Eur. J. Oper. Res.* **2016**, *250*, 615–627. [CrossRef]
60. Lai, D.S.W.; Demirag, O.C.; Leung, J.M.Y. A tabu search heuristic for the heterogeneous vehicle routing problem on a multigraph. *Trans. Res. E Logist. Trans. Rev.* **2016**, *86*, 32–52. [CrossRef]
61. Wu, Q.; Wang, Y.; Lü, Z. A tabu search based hybrid evolutionary algorithm for the max-cut problem. *Appl. Soft Comput.* **2015**, *34*, 827–837. [CrossRef]
62. Kvasnicka, V.; Pospichal, J. Fast evaluation of chemical distance by tabu search algorithm. *J. Chem. Inf. Comput. Sci.* **1994**, *34*, 1109–1112. [CrossRef]
63. Amaral, P.; Pais, T.C. Compromise Ratio with weighting functions in a Tabu Search multi-criteria approach to examination timetabling. *Comput. Oper. Res.* **2016**, *72*, 160–174. [CrossRef]
64. Shen, Q.; Shi, W.-M.; Kong, W. Modified tabu search approach for variable selection in quantitative structure–activity relationship studies of toxicity of aromatic compounds. *Artif. Intell. Med.* **2010**, *49*, 61–66. [CrossRef] [PubMed]
65. Chou, Y.H.; Yang, Y.J.; Chiu, C.H. Classical and quantum-inspired Tabu search for solving 0/1 knapsack problem. In Proceedings of the IEEE International Conference on Systems, Man, and Cybernetics (IEEE SMC 2011), Anchorage, AK, USA, 9–12 October 2011.
66. Huang, M.L. Hybridization of chaotic quantum particle swarm optimization with SVR in electric demand forecasting. *Energies* **2016**, *9*, 426. [CrossRef]
67. Amari, S.; Wu, S. Improving support vector machine classifiers by modifying kernel functions. *Neural Netw.* **1999**, *12*, 783–789. [CrossRef]
68. Chen, G.; Mao, Y.; Chui, C.K. Asymmetric image encryption scheme based on 3D chaotic cat maps. *Chaos Solitons Fractals* **2004**, *21*, 749–761. [CrossRef]
69. Su, H. Chaos quantum-behaved particle swarm optimization based neural networks for short-term load forecasting. *Procedia Eng.* **2011**, *15*, 199–203. [CrossRef]
70. 2014 Global Energy Forecasting Competition Site. Available online: http://www.drhongtao.com/gefcom/ (accessed on 2 May 2016).
71. Diebold, F.X.; Mariano, R.S. Comparing predictive accuracy. *J. Bus. Econ. Stat.* **1995**, *13*, 134–144.

energies

MDPI

Correction

Correction: Liang, Y., et al. Short-Term Load Forecasting Based on Wavelet Transform and Least Squares Support Vector Machine Optimized by Improved Cuckoo Search. *Energies* 2016, 9, 827

Yi Liang [1,*], Dongxiao Niu [1], Minquan Ye [2] and Wei-Chiang Hong [3]

[1] School of Economics and Management, North China Electric Power University, Beijing 102206, China;
niudx@126.com

[2] School of Economics and Management, North China Electric Power University, Baoding 070000, China;
hdymq2014@163.com

[3] Department of Information Management, Oriental Institute of Technology, New Taipei 220, Taiwan;
samuelsonhong@gmail.com

* Correspondence: lianglouis@126.com

Academic Editor: Enrico Sciubba
Received: 10 November 2016; Accepted: 9 December 2016; Published: 16 December 2016

The authors wish to make the following corrections to their paper [1]:

Please add the sentence "However, seasonality and long-term trends of the proposed model have not been tested and verified in this paper, which may become the limitation of this method, and the authors intend to study this aspect in the future." behind "so it can be applied widely in parameter optimization." in the Conclusion section.

The authors would like to apologize for any inconvenience caused to the readers by this change. The change does not affect the scientific results. The manuscript will be updated and the original will remain online on the article webpage.

Conflicts of Interest: The authors declare no conflict of interest.

Reference

1. Liang, Y.; Niu, D.X.; Ye, M.Q.; Hong, W.-C. Short-Term Load Forecasting Based on Wavelet Transform and Least Squares Support Vector Machine Optimized by Improved Cuckoo Search. *Energies* **2016**, *9*. [CrossRef]

MDPI

St. Alban-Anlage 66

4052 Basel

Switzerland

Tel. +41 61 683 77 34

Fax +41 61 302 89 18

www.mdpi.com

Energies Editorial Office

E-mail: energies@mdpi.com

www.mdpi.com/journal/energies

www.ingramcontent.com/pod-product-compliance
Lightning Source LLC
Chambersburg PA
CBHW051855210326
41597CB00033B/5911